Ecosystem-Based Mitigation: Harnessing Nature for Climate Solutions, Biodiversity Conservation, and Sustainable Development

1

Copyright

Global Climate Solutions

Table of Contents

Introduction

The Introduction sets the foundation for understanding ecosystem-based mitigation as a transformative approach to addressing climate change and environmental degradation. This chapter explores the urgency of mitigating greenhouse gas emissions, the critical role of ecosystems in achieving global climate goals, and the unique opportunities offered by nature-based solutions to create co-benefits for biodiversity, livelihoods, and resilience.

By examining the importance of integrating ecosystem-based approaches into climate strategies and development planning, this chapter highlights the interconnectedness of ecological health and socio-economic well-being. It also provides an overview of the book's structure, outlining key themes and strategies discussed in subsequent chapters. This context establishes a comprehensive framework for exploring the potential of ecosystems to drive sustainable and equitable climate action.

Importance of Ecosystem-Based Mitigation

Ecosystem-based mitigation is a critical approach to addressing the dual challenges of climate change and environmental degradation. This strategy leverages the natural processes of ecosystems to absorb and store carbon dioxide while simultaneously providing co-benefits such as biodiversity conservation, improved water security, and disaster risk reduction. By protecting, restoring, and sustainably managing ecosystems, ecosystem-based mitigation offers a cost-effective and scalable means of reducing greenhouse gas emissions.

Forests, wetlands, peatlands, soils, and marine systems are among the most significant natural carbon sinks. These ecosystems sequester vast amounts of carbon through photosynthesis, soil storage, and sediment deposition, playing a vital role in regulating the Earth's climate. For example, forests account for approximately 30% of global carbon storage, while peatlands store twice as much carbon as all the world's forests combined. These processes underscore the essential function ecosystems serve in maintaining atmospheric balance.

In addition to their climate mitigation potential, ecosystems provide critical services that benefit human well-being. Wetlands reduce flooding risks, forests supply clean water, and coastal ecosystems protect against storm surges. These co-benefits are particularly significant for vulnerable communities that depend heavily on natural resources for their livelihoods. Ecosystem-based mitigation thus aligns with global sustainability goals by addressing both environmental and social priorities.

However, the degradation of ecosystems due to deforestation, urbanization, and unsustainable agricultural practices undermines their ability to function as carbon sinks. These pressures not only increase greenhouse gas emissions but also diminish the ecosystems' capacity to provide essential services. Protecting and restoring these systems is therefore an urgent priority.

Ecosystem-based mitigation is also a pathway to achieving the objectives of international frameworks such as the Paris Agreement and the Sustainable Development Goals (SDGs). Its integration into climate strategies can enhance resilience, foster biodiversity, and promote sustainable development. While challenges such as funding, technical capacity, and policy alignment exist, the potential benefits far outweigh the obstacles.

In summary, ecosystem-based mitigation represents a vital, nature-centered approach to addressing the climate crisis. By maximizing the carbon sequestration potential of ecosystems and recognizing their broader value, this strategy can contribute to a more sustainable and resilient future for both people and the planet.

Role of Nature-Based Solutions in Climate Systems

Nature-based solutions (NbS) play a pivotal role in enhancing climate system resilience while addressing the pressing challenges of climate change. By utilizing natural processes and ecosystems, NbS contribute to both climate change mitigation and adaptation. These solutions encompass a wide range of practices, including reforestation, wetland restoration, sustainable agriculture, and coastal ecosystem conservation, each providing significant benefits for the climate and broader ecosystems.

One of the primary roles of NbS is their ability to sequester carbon dioxide from the atmosphere. Forests, mangroves, peatlands, and seagrass beds act as critical carbon sinks, absorbing and storing large quantities of carbon. For instance, mangroves and seagrass beds sequester carbon at rates up to ten times higher than terrestrial forests, demonstrating their importance in reducing atmospheric carbon levels. NbS also help regulate local climates by moderating temperatures, enhancing precipitation patterns, and mitigating the impacts of extreme weather events.

Beyond carbon sequestration, NbS strengthen the adaptive capacity of climate systems. Coastal ecosystems such as mangroves and salt marshes act as natural barriers, protecting communities from storm surges and sea-level rise. Similarly, forest ecosystems stabilize local weather patterns and reduce the risk of flooding. By maintaining biodiversity and ecosystem health, NbS enhance the ability of natural systems to adapt to changing climatic conditions, which in turn supports the resilience of human societies.

NbS also deliver a suite of co-benefits that align with global sustainability objectives. They improve water quality, support food security, and provide critical habitats for biodiversity. Importantly, NbS contribute to sustainable livelihoods, particularly for communities that rely on natural resources. By integrating social, economic, and environmental benefits, NbS provide holistic solutions to climate challenges.

However, implementing NbS at scale requires addressing challenges such as land-use conflicts, financing gaps, and inadequate governance frameworks. Effective implementation depends on integrating NbS into policy frameworks, fostering stakeholder collaboration, and ensuring community involvement. International initiatives like the UN Decade on Ecosystem Restoration highlight the growing recognition of NbS as essential tools for achieving global climate goals.

In conclusion, NbS are indispensable for supporting the stability and resilience of climate systems. By leveraging the power of nature, these solutions offer transformative potential to mitigate climate change impacts while promoting sustainable development and ecological health.

Overview of Book Structure and Chapter Synopsis

This book explores the potential of ecosystem-based mitigation as a sustainable solution to address climate change while delivering social, economic, and environmental co-benefits. Each chapter examines a critical component of ecosystem-based mitigation, offering practical insights and strategies for implementation.

Chapter 1 introduces the concept of ecosystem-based mitigation, defining its scope, potential, and co-benefits such as biodiversity conservation and disaster risk reduction. Chapters 2 through 6 focus on specific ecosystems—soils, agriculture, forests, peatlands, and marine systems—analyzing their roles in carbon sequestration and sustainable management practices. These chapters provide detailed discussions on techniques such as soil erosion control, sustainable farming, forest restoration, peatland conservation, and blue carbon solutions.

Chapter 7 highlights policy measures supporting ecosystem-based mitigation, including international frameworks, economic instruments, and multi-stakeholder approaches. Chapter 8 delves into tools and technologies for monitoring and measuring mitigation effectiveness, emphasizing advancements in remote sensing and data management.

Chapter 9 identifies common barriers to implementation, offering strategies for overcoming them through capacity building, public-private partnerships, and community engagement. Chapter 10 envisions the future of ecosystem-based mitigation, exploring emerging trends, innovations, and pathways for mainstreaming these solutions into development planning.

The book concludes with a call to action, emphasizing the importance of collaboration and innovation in achieving a sustainable, resilient future.

Chapter 1: Ecosystem-Based Mitigation and Its Co-Benefits

This chapter explores the concept of ecosystem-based mitigation, its definition, and its critical role in addressing the climate crisis. It highlights the immense climate mitigation potential of ecosystems, emphasizing their ability to sequester carbon and reduce greenhouse gas emissions. The chapter also delves into the wide-ranging co-benefits of ecosystem-based mitigation, including biodiversity conservation, enhanced livelihoods, water security, and disaster risk reduction, demonstrating its alignment with global sustainability goals.

Additionally, the challenges associated with implementing ecosystem-based mitigation strategies are examined, including financial, technical, and institutional barriers. By presenting a comprehensive overview, this chapter lays the groundwork for understanding how ecosystem-based mitigation can serve as a transformative approach to achieving climate and sustainable development objectives.

Definition and Scope of Ecosystem-Based Mitigation

Ecosystem-based mitigation refers to the strategic use of ecosystems to address climate change by enhancing their natural ability to sequester carbon and reduce greenhouse gas emissions. This approach leverages the inherent processes of ecosystems, such as carbon absorption and storage, to contribute to global climate goals. By prioritizing the protection, restoration, and sustainable management of ecosystems, ecosystem-based mitigation provides a nature-centered solution to the climate crisis.

The scope of ecosystem-based mitigation encompasses a wide range of ecosystems, including forests, peatlands, wetlands, grasslands, soils, and marine environments. These ecosystems act as vital carbon sinks, absorbing carbon dioxide from the atmosphere and storing it in biomass, soil, and sediments. For instance, forests are responsible for storing approximately 45% of terrestrial carbon, while peatlands hold twice as much carbon as all the world's forests combined. Coastal and marine ecosystems, such as mangroves and seagrass beds, also play significant roles in sequestering carbon and mitigating the effects of climate change.

In addition to mitigating climate change, ecosystem-based approaches provide multiple co-benefits, including biodiversity conservation, enhanced water security, disaster risk reduction, and improved livelihoods for local communities. These benefits highlight the interconnectedness between climate systems, ecosystems, and human well-being, making ecosystem-based mitigation a holistic and inclusive strategy.

The scope of this approach also extends to policy and governance, where ecosystem-based mitigation is increasingly integrated into international climate agreements, such as the Paris Agreement, and national climate strategies. These policies aim to incentivize ecosystem protection and restoration through mechanisms like carbon pricing, payments for ecosystem services, and reforestation initiatives. Furthermore, ecosystem-based mitigation aligns closely with the United Nations SDGs, addressing objectives related to climate action, biodiversity, and sustainable development.

However, the implementation of ecosystem-based mitigation faces challenges, including land-use conflicts, limited financial resources, and insufficient technical capacity. Addressing these barriers requires multi-stakeholder collaboration, innovative funding mechanisms, and robust monitoring systems to ensure effective outcomes.

In summary, ecosystem-based mitigation is a crucial, nature-centered solution to climate change. Its comprehensive scope addresses environmental, social, and economic priorities, offering a pathway to a more sustainable and resilient future for both people and the planet.

Climate Mitigation Potential of Ecosystems

Ecosystems play a pivotal role in mitigating climate change by acting as natural carbon sinks, absorbing and storing significant amounts of carbon dioxide from the atmosphere. Forests, wetlands, peatlands, grasslands, soils, and marine ecosystems are vital to maintaining the Earth's carbon balance. Their ability to sequester carbon, combined with their contribution to reducing greenhouse gas emissions, makes them indispensable in global efforts to combat climate change.

Forests are among the most effective ecosystems for carbon sequestration. Through photosynthesis, trees absorb carbon dioxide, storing it in their biomass and soil. Tropical, temperate, and boreal forests collectively store approximately 45% of terrestrial carbon. Reforestation, afforestation, and sustainable forest management enhance this sequestration capacity, allowing forests to absorb additional carbon while maintaining their ecological integrity.

Peatlands, though covering only 3% of the Earth's land surface, store twice as much carbon as all forests combined. These waterlogged ecosystems accumulate organic material over millennia, creating an immense carbon reservoir. However, when disturbed through drainage, agriculture, or fires, peatlands become significant sources of carbon emissions. Restoring degraded peatlands and preventing further exploitation are crucial to preserving their carbon storage capacity.

Grasslands and savannas also play an important role in carbon sequestration, particularly in their soil organic matter. These ecosystems store carbon below ground, offering a stable and long-term carbon sink. Sustainable grazing practices and the prevention of land degradation are essential for maintaining their sequestration potential.

Soils, across all terrestrial ecosystems, are another critical component of the global carbon cycle. Agricultural soils, when managed sustainably, can sequester significant amounts of carbon. Practices like reduced tillage, cover cropping, and agroforestry enhance soil carbon storage while improving soil health and productivity. The restoration of degraded lands also contributes to carbon sequestration, making soils a key element in ecosystem-based mitigation strategies.

Coastal and marine ecosystems, including mangroves, seagrass meadows, and tidal salt marshes, provide a unique form of carbon storage known as "blue carbon." These ecosystems sequester carbon at rates up to ten times higher than terrestrial forests, storing it in plant biomass and underwater sediments. Protecting and restoring these ecosystems is essential for maintaining their role as effective carbon sinks.

While the climate mitigation potential of ecosystems is immense, their effectiveness is vulnerable to human activities such as deforestation, land conversion, and pollution. Degraded ecosystems not only lose their carbon storage capacity but also become sources of greenhouse gas emissions. Preventing degradation and implementing restoration efforts are vital to unlocking the full potential of ecosystems as climate solutions.

In conclusion, ecosystems offer a powerful, nature-based pathway for mitigating climate change. By enhancing the protection, restoration, and sustainable management of forests, soils, peatlands, and marine environments, we can harness their ability to sequester carbon while delivering additional ecological and socio-economic benefits. These strategies represent a critical component of achieving global climate targets and building a more resilient future.

Exploring Co-Benefits: Biodiversity Conservation, Improved Livelihoods, Water Security, and Disaster Risk Reduction

Ecosystem-based mitigation offers a unique advantage by simultaneously addressing climate change while delivering a range of co-benefits that enhance ecological health, societal well-being, and economic stability. These co-benefits underscore the interconnectedness of natural systems and human life, making ecosystem-based approaches a holistic and impactful strategy for sustainable development. Among the most significant co-benefits are biodiversity conservation, improved livelihoods, water security, and disaster risk reduction.

Biodiversity Conservation

Ecosystems play a critical role in maintaining global biodiversity, which is essential for ecological balance and resilience. By protecting and restoring ecosystems such as forests, wetlands, and coral reefs, ecosystem-based mitigation safeguards habitats for countless species. Forest conservation, for example, protects diverse flora and fauna, many of which are endemic and at risk of extinction. Similarly, restoring degraded coastal ecosystems like mangroves and seagrass meadows provides vital habitats for marine life. Biodiversity also supports ecosystem services, such as pollination and nutrient cycling, which are integral to agricultural productivity and food security. Through these efforts, ecosystem-based mitigation contributes to halting biodiversity loss and fostering ecological resilience in the face of climate change.

Improved Livelihoods

Ecosystem-based mitigation has profound socio-economic benefits, particularly for communities that depend on natural resources for their livelihoods. Sustainable forest management, for instance, provides employment opportunities in reforestation, agroforestry, and non-timber forest product harvesting. Similarly, restoring wetlands supports fisheries and ecotourism, generating income for local populations. Agricultural practices that enhance soil health and carbon sequestration, such as agroforestry and crop diversification, increase productivity and income for farmers. These initiatives also promote social equity by empowering marginalized communities, including women and Indigenous peoples, who are often most affected by environmental degradation and climate change. By integrating climate action with economic opportunities, ecosystem-based mitigation enhances the resilience and prosperity of vulnerable populations.

Water Security

Healthy ecosystems are vital for ensuring water availability and quality. Forests, wetlands, and soils regulate the water cycle by enhancing groundwater recharge, maintaining river flows, and improving water filtration. For example, forested watersheds act as natural reservoirs, storing rainwater and releasing it gradually to sustain rivers and streams. Wetlands filter pollutants, reducing contamination and improving water quality for downstream users. Sustainable agricultural practices, such as contour farming and reduced tillage, prevent soil erosion and sedimentation, protecting water bodies from degradation. Ecosystem-based mitigation thus addresses water scarcity and quality challenges, supporting communities and industries dependent on reliable water supplies.

Disaster Risk Reduction

Ecosystems serve as natural buffers against climate-related disasters, reducing risks and enhancing community resilience. Coastal ecosystems, such as mangroves and salt marshes, protect shorelines from storm surges, sea-level rise, and erosion. These ecosystems dissipate wave energy, reducing the impact of extreme weather events on coastal communities and infrastructure. Similarly, forests on slopes stabilize soils, preventing landslides and mitigating the impact of heavy rainfall. Wetlands act as natural sponges, absorbing excess water and reducing flood risks. By maintaining and restoring these ecosystems, ecosystem-based mitigation reduces the frequency and severity of natural disasters, lowering economic losses and protecting lives.

Synergies Between Co-Benefits

The co-benefits of ecosystem-based mitigation are deeply interconnected, creating synergies that amplify their impact. For example, conserving biodiversity enhances ecosystem resilience, enabling natural systems to adapt to changing climatic conditions and continue providing essential services. Improved water security, in turn, supports livelihoods and reduces vulnerabilities to climate-related shocks. By addressing multiple objectives simultaneously, ecosystem-based mitigation strengthens the foundation for achieving global sustainability goals, including the United Nations SDGs.

Challenges and Opportunities

While the co-benefits of ecosystem-based mitigation are significant, realizing them requires addressing various challenges. Competing land-use demands, insufficient funding, and lack of technical capacity often hinder the implementation of ecosystem-based approaches. Overcoming these barriers involves fostering collaboration among governments, communities, and private sectors, as well as developing innovative financing mechanisms, such as payments for ecosystem services. Monitoring and evaluating the co-benefits of ecosystem-based mitigation is also essential to ensure long-term success and equitable outcomes.

Challenges in Implementing Ecosystem-Based Mitigation Strategies

Ecosystem-based mitigation strategies offer significant potential to address climate change, but their implementation is often hindered by a range of challenges. These obstacles, which span financial, institutional, technical, and social dimensions, can limit the scalability and effectiveness of such approaches if not addressed systematically.

Financial Barriers

Securing adequate and sustained funding is a major challenge for ecosystem-based mitigation. The upfront costs of restoration projects, conservation programs, and sustainable management practices can be high, particularly in resource-constrained settings. Additionally, financial mechanisms such as payments for ecosystem services and carbon markets remain underutilized or inaccessible in many regions. The lack of long-term investment hinders the ability to maintain and scale these initiatives.

Institutional and Policy Challenges

Weak governance structures and misaligned policies can undermine the implementation of ecosystem-based mitigation. In many cases, land-use planning, agricultural policies, and development priorities conflict with conservation objectives. Overlapping jurisdictional responsibilities and limited coordination among government agencies further complicate efforts. Additionally, the absence of clear legal frameworks to support ecosystem protection and restoration limits the effectiveness of such strategies.

Technical Constraints

Technical capacity gaps, including insufficient knowledge and expertise, pose significant challenges. Many stakeholders lack access to scientific data, tools, and methodologies required for designing and monitoring ecosystem-based mitigation projects. This limitation is particularly acute in developing countries, where resources for capacity building and training are often scarce. Furthermore, the complexity of ecosystem dynamics and climate interactions requires multidisciplinary expertise, which can be difficult to mobilize.

Social and Cultural Barriers

The success of ecosystem-based mitigation depends on local community engagement and support. However, social and cultural factors can impede participation. In some cases, communities may lack awareness of the benefits of ecosystem-based approaches or perceive them as conflicting with traditional practices and livelihoods. Land tenure issues and inequitable benefit-sharing mechanisms can also create resistance, particularly among marginalized groups.

Chapter 2: Soil: Erosion Control and Vegetative Measures

Soil plays a fundamental role in supporting life on Earth, acting as a vital resource for food production, water filtration, and carbon storage. This chapter delves into the importance of soil in ecosystem-based mitigation, focusing on its potential for carbon sequestration and its role in combating climate change. It explores various strategies for controlling soil erosion and maintaining soil health, with an emphasis on sustainable techniques such as terracing, contour farming, and the use of vegetative measures.

The chapter also highlights the benefits of practices like agroforestry, cover crops, and grass strips in enhancing soil stability and fertility. Finally, it examines the role of policies and incentives in promoting soil conservation, providing a comprehensive overview of how soil management contributes to global climate goals and sustainable development. This chapter underscores soil's critical role in ecosystem health and resilience, offering actionable solutions for addressing erosion and improving soil quality.

The Role of Soil in Carbon Sequestration

Soil is a critical component of the Earth's carbon cycle and plays a vital role in mitigating climate change through carbon sequestration. By storing carbon in organic and inorganic forms, soils act as a significant carbon sink, capable of offsetting a substantial portion of global greenhouse gas emissions. The ability of soils to absorb and store carbon makes them an integral part of ecosystem-based mitigation strategies.

Soil organic carbon (SOC) is the primary form of carbon storage in soils, derived from decomposed plant and animal matter. This organic matter not only enriches soil fertility but also contributes to long-term carbon storage by forming stable compounds that resist decomposition. Soil inorganic carbon, which includes carbonate minerals, provides additional carbon storage, particularly in arid and semi-arid regions. Together, these carbon pools represent a significant portion of terrestrial carbon storage, with estimates suggesting that soils hold more carbon than the atmosphere and vegetation combined.

Agricultural soils offer considerable potential for enhanced carbon sequestration through sustainable management practices. Techniques such as reduced tillage, cover cropping, and crop rotation minimize soil disturbance and increase organic matter inputs, leading to greater carbon retention. Agroforestry systems, which integrate trees into agricultural landscapes, further enhance carbon sequestration by combining above-ground biomass with below-ground soil carbon storage. These practices not only improve carbon storage but also enhance soil health and productivity, providing multiple benefits.

Grasslands and rangelands are another critical area for soil carbon sequestration. Through proper grazing management and restoration of degraded lands, these ecosystems can accumulate significant amounts of organic carbon in their soils. Similarly, reforestation and afforestation efforts contribute to soil carbon sequestration by increasing organic matter inputs from leaf litter and root systems.

The potential of soils to sequester carbon is influenced by several factors, including soil type, climate, land use, and management practices. For instance, soils in colder climates tend to store more carbon due to slower decomposition rates, while tropical soils may experience higher carbon losses if improperly managed. Restoring degraded soils, particularly those affected by erosion, compaction, and nutrient depletion, is essential for maximizing their carbon sequestration capacity.

Despite their potential, soils are vulnerable to degradation from unsustainable practices such as deforestation, overgrazing, and intensive agriculture. These activities reduce soil organic matter, disrupt soil structure, and release stored carbon into the atmosphere. Protecting and restoring soils through sustainable land management practices is therefore crucial for maintaining their role as a carbon sink.

Techniques to Control Soil Erosion, Such as Terracing and Contour Farming

Soil erosion is a significant environmental challenge that threatens agricultural productivity, water quality, and ecosystem health. It involves the removal of topsoil, which contains essential nutrients and organic matter, by natural forces such as wind and water. Controlling soil erosion is a critical component of sustainable land management and ecosystem-based mitigation, as it helps maintain soil fertility, reduce sedimentation in water bodies, and protect the soil's carbon storage capacity. Among the most effective techniques for soil erosion control are terracing and contour farming, both of which use natural landscape features to minimize soil loss.

Terracing

Terracing is a widely used technique to control soil erosion on sloped land. It involves the creation of stepped, leveled areas across a slope to slow down water flow and reduce soil loss. Each terrace acts as a barrier, capturing water and sediment, which would otherwise be washed downhill. This technique is particularly effective in regions with heavy rainfall or steep terrains, where erosion risks are high.

Terracing also contributes to water conservation by allowing rainwater to infiltrate the soil instead of running off the surface. This increased water retention benefits crops, reducing the need for irrigation in rain-fed agricultural systems. Additionally, terraces improve the distribution of organic matter and nutrients, enhancing soil fertility and crop yields. While constructing terraces requires substantial initial labor and investment, their long-term benefits in preventing erosion and supporting sustainable agriculture outweigh the costs.

Contour Farming

Contour farming is another effective technique for controlling soil erosion, especially on gently sloping lands. This practice involves plowing and planting along the natural contours of the land rather than up and down the slope. By following the land's natural shape, contour farming creates furrows and ridges that act as barriers to water flow, reducing runoff and soil erosion.

This method is particularly beneficial in regions with moderate rainfall, where it helps retain moisture in the soil, improving crop growth. Contour farming also minimizes the loss of fertilizers and pesticides, which might otherwise be carried away by runoff, thereby reducing pollution in nearby water bodies. The simplicity and cost-effectiveness of contour farming make it an accessible option for smallholder farmers, contributing to its widespread adoption in various parts of the world.

Other Erosion Control Techniques

In addition to terracing and contour farming, several complementary practices can enhance soil stability and reduce erosion. Strip cropping, for instance, involves alternating rows of different crops, such as grains and legumes, to break the flow of water and wind across the field. This practice also improves soil fertility and crop diversity, promoting sustainable farming systems.

Buffer strips, composed of grasses or shrubs, can be planted along the edges of fields, streams, or waterways to trap sediment and slow down runoff. These vegetative buffers act as a natural filter, protecting water bodies from sedimentation and agricultural pollutants.

Windbreaks, or shelterbelts, are rows of trees or shrubs planted perpendicular to prevailing winds to reduce wind speed and prevent soil erosion. These barriers are particularly effective in arid and semi-arid regions, where wind erosion poses a significant threat to agricultural land.

Reduced tillage and no-till farming practices also play a critical role in minimizing soil disturbance and preventing erosion. By leaving crop residues on the soil surface, these methods protect the soil from the impact of rain and wind, enhancing organic matter retention and improving soil structure over time.

Challenges and Considerations

While these techniques are effective, their implementation depends on various factors, including land characteristics, climate, and socio-economic conditions. Terracing, for example, requires significant labor and investment, which may not be feasible for small-scale farmers without external support. Similarly, the effectiveness of contour farming depends on proper planning and maintenance to prevent waterlogging and soil compaction.

Integrating these erosion control techniques into broader land management practices can maximize their benefits. For instance, combining terracing or contour farming with agroforestry or cover cropping enhances soil stability while providing additional benefits such as biodiversity conservation and carbon sequestration.

Vegetative Measures: Cover Crops, Agroforestry, and Grass Strips

Vegetative measures are essential tools for controlling soil erosion, enhancing soil fertility, and contributing to ecosystem-based mitigation strategies. By utilizing plants and vegetation cover, these measures protect soil from degradation while offering additional benefits such as carbon sequestration, biodiversity conservation, and improved water quality. Among the most effective vegetative measures are cover crops, agroforestry, and grass strips, which can be implemented in diverse agricultural landscapes to promote sustainable land management.

Cover Crops

Cover crops are plants grown primarily to protect and improve soil health rather than for harvest. These crops, such as legumes, grasses, and cereals, provide a protective layer that reduces soil erosion caused by wind and water. By maintaining continuous ground cover, cover crops minimize the impact of raindrops on the soil surface, preventing crust formation and runoff.

In addition to erosion control, cover crops contribute to soil fertility by adding organic matter and enhancing soil structure. Leguminous cover crops, such as clover and vetch, fix atmospheric nitrogen, enriching the soil with this vital nutrient for subsequent crops. The decomposition of cover crop residues increases soil organic carbon levels, which improves water retention and promotes microbial activity. This process also enhances the soil's carbon sequestration potential, making cover crops a valuable component of climate mitigation strategies.

Cover crops also suppress weeds, reduce the need for chemical herbicides, and break pest and disease cycles. They are particularly effective in no-till or reduced-tillage systems, where their roots stabilize the soil, and their residues act as mulch. Implementing cover crops requires careful selection based on local climate, soil type, and cropping systems to maximize their benefits.

Agroforestry

Agroforestry integrates trees and shrubs into agricultural systems, creating a multifunctional landscape that supports sustainable land management and ecosystem resilience. This practice enhances soil protection by reducing erosion, as tree roots stabilize the soil and intercept rainfall. The canopy of trees minimizes the impact of heavy rains, while fallen leaves contribute organic matter to the soil, improving its structure and fertility.

Agroforestry systems sequester significant amounts of carbon, both above ground in tree biomass and below ground in soil organic matter. These systems also diversify income sources for farmers by providing products such as fruits, nuts, timber, and medicinal plants. For example, alley cropping, a type of agroforestry, involves planting rows of trees between crops, optimizing land use and enhancing biodiversity.

Silvopasture, another agroforestry practice, combines trees with livestock grazing. This approach reduces soil compaction and erosion caused by livestock while providing shade and shelter for animals. Agroforestry also improves water management by enhancing groundwater recharge and reducing surface runoff. Despite its numerous benefits, agroforestry requires long-term planning and investment, as trees take time to mature and deliver measurable outcomes.

Grass Strips

Grass strips are narrow bands of perennial grasses planted along the contour of fields or at intervals across slopes to trap sediment, reduce runoff, and stabilize soil. These strips act as natural barriers, slowing down water flow and allowing suspended particles to settle, thereby preventing soil erosion. Grass strips are particularly effective on sloped agricultural lands, where they intercept water moving downhill and minimize soil loss.

The choice of grasses for these strips depends on local conditions, with species like vetiver, switchgrass, and Bermuda grass commonly used due to their deep roots and adaptability. Grass strips also contribute to biodiversity by providing habitats for pollinators and other beneficial organisms. Additionally, they improve water infiltration, reducing the risk of surface runoff and enhancing groundwater recharge.

In areas prone to heavy rainfall, grass strips act as buffers that protect adjacent water bodies from sedimentation and nutrient runoff. This function is particularly important for maintaining water quality in agricultural regions. Grass strips are relatively low-cost and easy to establish, making them accessible to smallholder farmers. However, maintaining these strips requires periodic cutting to prevent them from becoming overly dense and to recycle nutrients back into the soil.

Synergies Between Vegetative Measures

The combined use of cover crops, agroforestry, and grass strips can create synergistic effects that enhance the overall effectiveness of soil conservation and ecosystem-based mitigation. For example, integrating agroforestry systems with cover crops provides both above-ground and below-ground protection against erosion while improving soil health and carbon storage. Similarly, grass strips can be used in conjunction with agroforestry to stabilize slopes and protect water bodies.

Challenges and Considerations

Implementing vegetative measures requires careful planning, as their effectiveness depends on local conditions, including soil type, climate, and land use. Farmers may face challenges such as high initial labor demands, competition for land, and the need for technical knowledge to design and maintain these systems effectively. Providing incentives, training, and access to resources can help overcome these barriers and promote widespread adoption.

Policies and Incentives for Promoting Soil Conservation Practices

Soil conservation is a cornerstone of sustainable land management, yet its implementation often requires supportive policies and incentives to encourage widespread adoption. Governments, organizations, and international bodies have developed various mechanisms to promote soil conservation, addressing the financial, technical, and social barriers faced by land managers and farmers.

These policies and incentives are crucial for ensuring the long-term health of soils, maintaining ecosystem services, and advancing climate mitigation goals.

Subsidies and Financial Incentives

Financial support is one of the most effective tools for promoting soil conservation practices. Subsidies for sustainable agricultural methods, such as reduced tillage, cover cropping, and agroforestry, lower the initial costs for farmers and encourage adoption. Payments for ecosystem services (PES) schemes reward landowners for maintaining practices that enhance soil health and sequester carbon. For example, farmers may receive compensation for planting cover crops or restoring degraded land, recognizing the broader environmental benefits of these actions.

Carbon markets also provide financial incentives by allowing farmers to generate carbon credits for practices that increase soil carbon sequestration. These credits can be traded in voluntary or compliance markets, offering an additional income stream while contributing to global emissions reduction targets. However, the accessibility of carbon markets needs to be expanded, particularly for small-scale farmers.

Regulatory Frameworks

Clear and enforceable regulations play a critical role in preventing soil degradation. Policies that restrict harmful activities, such as deforestation, overgrazing, and unsustainable agricultural practices, are essential for protecting soil resources. Land-use planning regulations can promote sustainable practices by encouraging crop diversification, erosion control measures, and buffer zones near water bodies.

Mandatory soil health assessments and monitoring can also be integrated into regulatory frameworks, ensuring compliance with conservation standards. For instance, governments may require farmers to meet specific criteria related to soil organic matter levels, erosion rates, or nutrient management plans. These regulations ensure accountability while fostering a culture of sustainable land management.

Incentives for Research and Education

Investments in research and education are crucial for advancing soil conservation practices. Policies that fund research into innovative techniques, such as biochar application, microbial inoculants, and precision agriculture, contribute to the development of more effective and efficient soil management strategies. Educational initiatives, including farmer training programs and extension services, provide the knowledge and skills needed to implement conservation practices successfully.

Community-Based Approaches

Community engagement is essential for the success of soil conservation policies. Incentives that promote collaborative efforts, such as watershed management programs or community-led reforestation projects, strengthen local ownership and accountability. These initiatives often involve shared benefits, such as improved water quality or increased agricultural productivity, which further motivate participation.

International Support and Collaboration

Global frameworks, such as the United Nations' SDGs, emphasize the importance of soil conservation in achieving sustainability. International funding mechanisms, including grants from organizations like the Global Environment Facility (GEF) and the Green Climate Fund (GCF), provide critical resources for implementing soil conservation projects in developing countries. Collaborative efforts between governments, non-governmental organizations, and private sectors ensure that resources and expertise are effectively leveraged.

Chapter 3: Agriculture: Sustainable Agricultural Techniques

Agriculture lies at the heart of human sustenance and economic development, yet it is also a significant contributor to greenhouse gas emissions and environmental degradation. This chapter explores the transformative role of sustainable agricultural techniques in addressing these challenges. It examines how carbon sequestration in agricultural systems can mitigate climate change while enhancing soil health and productivity.

The chapter delves into practices such as crop rotation, reduced tillage, and organic farming, highlighting their environmental and economic benefits. It also emphasizes the integration of trees into agricultural landscapes through agroforestry, showcasing its potential to balance productivity with ecosystem conservation. Finally, the chapter discusses strategies to manage agricultural emissions, including improved manure management and efficient irrigation, presenting a holistic approach to fostering sustainable and resilient agricultural systems. By adopting these practices, agriculture can become a pivotal force in achieving global sustainability and climate goals.

Carbon Sequestration in Agricultural Systems

Agricultural systems play a critical role in mitigating climate change by serving as significant carbon sinks. Through sustainable practices, these systems can capture and store atmospheric carbon dioxide in soil organic matter and plant biomass, contributing to reduced greenhouse gas concentrations. This process, known as carbon sequestration, enhances the agricultural sector's potential to align productivity with climate mitigation goals.

SOC Storage

The soil is the largest terrestrial carbon reservoir, and sustainable agricultural practices can significantly enhance its carbon storage capacity. Techniques such as reduced tillage, cover cropping, and crop rotation minimize soil disturbance and increase organic matter inputs, leading to higher SOC levels. Cover crops, for instance, prevent soil erosion and contribute to carbon storage by adding plant residues to the soil. Crop rotation diversifies organic inputs, enriching the soil and improving its ability to retain carbon.

Agroforestry, which integrates trees with crops and livestock, further boosts carbon sequestration by combining above-ground and below-ground carbon storage. Tree roots contribute to soil structure, reducing erosion and promoting long-term carbon stability. The trees themselves act as biomass carbon sinks, capturing CO_2 during photosynthesis and storing it in their trunks, branches, and leaves.

Carbon Sequestration in Perennial Crops

Perennial crops, such as fruit trees and shrubs, have a unique advantage in sequestering carbon. Unlike annual crops that require replanting, perennial plants establish long-term root systems that continuously add organic matter to the soil. Their extensive root networks stabilize soil, reduce erosion, and enhance water retention, creating optimal conditions for carbon storage.

Livestock Management and Carbon Sequestration

Sustainable grazing practices also contribute to carbon sequestration. Well-managed grazing systems promote healthy grasslands, which are effective at storing carbon in soil organic matter. Rotational grazing, which allows pastures to recover between grazing periods, prevents overgrazing and soil degradation, enhancing carbon storage. Silvopasture, a practice that integrates trees with grazing areas, provides additional benefits by sequestering carbon in both trees and soil.

Challenges and Considerations

Despite its potential, carbon sequestration in agricultural systems faces several challenges. Degraded soils, land-use changes, and unsustainable practices can reduce the effectiveness of sequestration efforts. Additionally, the permanence of stored carbon is not guaranteed, as improper management or disturbances, such as soil erosion, can release carbon back into the atmosphere. Addressing these challenges requires long-term planning, monitoring, and adherence to sustainable practices.

Policy and Incentives

Supportive policies and financial incentives are critical for scaling carbon sequestration in agriculture. Initiatives such as carbon credits, payments for ecosystem services, and subsidies for sustainable farming practices encourage adoption and ensure that farmers receive economic benefits for their efforts. International frameworks, like the United Nations SDGs, further promote these practices by emphasizing their role in climate action and sustainable land management.

Sustainable Farming Practices: Crop Rotation, Reduced Tillage, and Organic Farming

Sustainable farming practices are essential to ensuring long-term agricultural productivity, environmental health, and climate resilience. These practices aim to optimize resource use, reduce environmental impacts, and promote biodiversity while maintaining or improving yields. Among the most effective approaches are crop rotation, reduced tillage, and organic farming, each of which offers unique benefits for soil health, carbon sequestration, and sustainable land management.

Crop Rotation

Crop rotation involves the systematic planting of different crops in a sequential pattern on the same land over successive growing seasons. This practice enhances soil fertility, disrupts pest and disease cycles, and reduces the need for chemical inputs like fertilizers and pesticides. By alternating crops with varying nutrient demands and root structures, crop rotation prevents nutrient depletion and promotes soil health.

For example, including legumes in a rotation can fix atmospheric nitrogen into the soil, reducing the need for synthetic nitrogen fertilizers. Deep-rooted crops, such as sunflowers or alfalfa, improve soil structure by breaking up compacted layers and enhancing water infiltration. Additionally, crop rotation minimizes the risk of soil erosion, as a diverse plant cover stabilizes the soil and reduces surface runoff. These benefits collectively contribute to the long-term sustainability of agricultural systems.

Crop rotation also supports carbon sequestration by increasing organic matter inputs to the soil. The residues from diverse crops enrich the soil, fostering microbial activity and improving carbon storage. This practice is particularly effective when integrated with cover cropping and other soil conservation measures.

Reduced Tillage

Reduced tillage, also known as conservation tillage, minimizes soil disturbance by limiting or eliminating plowing and other mechanical operations. Traditional tillage practices often lead to soil erosion, compaction, and the release of stored carbon into the atmosphere. In contrast, reduced tillage preserves soil structure, retains organic matter, and enhances the soil's capacity to sequester carbon.

By leaving crop residues on the surface, reduced tillage protects the soil from the impact of wind and water erosion. These residues act as a natural mulch, conserving soil moisture and suppressing weeds. The practice also fosters the development of a stable soil ecosystem, promoting beneficial microorganisms and improving nutrient cycling.

Adopting reduced tillage requires adjustments in weed management and planting techniques, as the undisturbed soil surface may initially harbor weed growth. However, combining reduced tillage with other sustainable practices, such as crop rotation and cover cropping, can mitigate these challenges and optimize its benefits.

Organic Farming

Organic farming is a holistic approach to agriculture that relies on natural inputs and ecological processes to enhance soil fertility, promote biodiversity, and reduce environmental impacts. This practice avoids synthetic fertilizers, pesticides, and genetically modified organisms (GMOs), instead emphasizing the use of organic materials, crop diversification, and biological pest control.

Composting is a cornerstone of organic farming, as it recycles organic waste into nutrient-rich soil amendments. The application of compost improves soil structure, enhances water retention, and increases organic carbon content, making it a key contributor to carbon sequestration. Organic farming also employs green manures and cover crops to enrich the soil and protect it from erosion.

Crop diversity is another critical element of organic farming. By planting a variety of crops, organic farmers create a resilient agroecosystem that reduces pest outbreaks and supports beneficial organisms. This diversity fosters ecological balance and minimizes the need for chemical interventions.

While organic farming offers numerous environmental and health benefits, it requires careful management to maintain productivity and economic viability. The transition to organic methods may involve higher labor costs and lower initial yields, but over time, the improved soil health and ecosystem resilience can offset these challenges.

Benefits and Synergies

The integration of crop rotation, reduced tillage, and organic farming creates synergies that amplify their individual benefits. For instance, combining reduced tillage with organic farming enhances soil carbon sequestration by preserving organic matter and minimizing soil disturbance. Similarly, incorporating crop rotation into organic systems diversifies nutrient sources and strengthens pest management, reducing reliance on external inputs.

Challenges and Considerations

Despite their advantages, sustainable farming practices face barriers to widespread adoption. These include financial constraints, lack of technical knowledge, and limited access to resources such as compost or cover crop seeds. Addressing these challenges requires supportive policies, training programs, and incentives that empower farmers to transition to sustainable methods.

Role of Agroforestry in Integrating Trees into Agricultural Landscapes

Agroforestry is a sustainable land management practice that integrates trees and shrubs into agricultural landscapes. By combining agricultural crops, livestock, and forestry elements on the same land, agroforestry creates multifunctional systems that offer environmental, economic, and social benefits. This practice is increasingly recognized for its role in addressing climate change, enhancing biodiversity, and improving agricultural productivity.

Enhancing Carbon Sequestration

One of the most significant contributions of agroforestry is its ability to sequester carbon. Trees and shrubs absorb carbon dioxide from the atmosphere during photosynthesis and store it in their biomass and root systems. The organic matter from fallen leaves, branches, and roots enriches the soil, increasing soil organic carbon levels. This dual sequestration capacity—above and below ground—makes agroforestry a powerful tool for climate mitigation.

In addition, agroforestry systems protect soil carbon by reducing erosion and maintaining soil structure. For example, the presence of tree roots stabilizes soil, preventing the loss of organic matter that contributes to carbon storage. These attributes highlight the critical role of agroforestry in integrating carbon sequestration into agricultural systems.

Improving Soil Health and Fertility

Agroforestry enhances soil health by increasing organic matter, improving nutrient cycling, and promoting biodiversity in the soil ecosystem. Tree roots penetrate deep into the soil, bringing up nutrients from lower layers that are otherwise inaccessible to shallow-rooted crops. The decomposition of tree leaves and branches adds organic matter, enriching the soil and supporting microbial activity.

Nitrogen-fixing trees, such as those in the legume family, improve soil fertility by converting atmospheric nitrogen into forms usable by plants. This natural fertilization reduces the need for synthetic fertilizers, lowering costs for farmers and minimizing the environmental impacts of chemical inputs. The improved soil structure and fertility provided by agroforestry systems also increase water infiltration and retention, reducing the risk of drought stress on crops.

Biodiversity Conservation

Agroforestry promotes biodiversity by creating diverse habitats for wildlife, insects, and soil organisms. The incorporation of trees and shrubs into agricultural landscapes provides food and shelter for pollinators, which are essential for crop production. Birds and other wildlife benefit from the increased vegetation cover, contributing to ecological balance.

Biodiversity within agroforestry systems also enhances pest and disease management. Predatory insects and birds that inhabit these systems help control pests, reducing the need for chemical pesticides. This natural pest regulation improves crop yields and promotes a healthier environment.

Economic Benefits for Farmers

Agroforestry diversifies income sources for farmers, making agricultural systems more resilient to economic and climatic shocks. Trees and shrubs in agroforestry systems provide timber, fruits, nuts, medicinal plants, and other products that can be sold in local and global markets. For example, alley cropping—where rows of trees are planted alongside crops—allows farmers to harvest tree products while maintaining traditional agricultural production.

Silvopasture, which integrates trees with livestock grazing, is another agroforestry practice that enhances productivity. Trees provide shade and shelter for animals, improving their welfare and reducing heat stress. The integration of multiple revenue streams reduces the risk of income loss due to crop failure or market fluctuations.

Climate Adaptation and Resilience

Agroforestry systems improve the resilience of agricultural landscapes to climate change. Trees act as windbreaks, protecting crops from strong winds and reducing soil erosion. They also moderate microclimates by reducing temperature extremes and increasing humidity, creating favorable conditions for crop growth. These benefits are particularly important in regions prone to droughts, floods, or other climatic challenges.

Agroforestry also supports water management by enhancing groundwater recharge and reducing surface runoff. The improved water availability benefits crops and reduces the risk of waterlogging or flooding.

Managing Agricultural Emissions Through Improved Manure Management and Efficient Irrigation

Agriculture is a significant source of greenhouse gas (GHG) emissions, particularly methane (CH_4) and nitrous oxide (N_2O), which have high global warming potential. Improved manure management and efficient irrigation are critical strategies for mitigating these emissions while enhancing agricultural productivity and sustainability. By addressing emissions from livestock waste and nitrogen-intensive practices, these approaches contribute to ecosystem-based mitigation and climate resilience.

Improved Manure Management

Livestock manure is a major source of methane and nitrous oxide emissions, primarily from anaerobic decomposition and the volatilization of nitrogen compounds. Proper management of manure can significantly reduce these emissions while providing a valuable resource for soil fertility and carbon sequestration.

Anaerobic Digestion

Anaerobic digestion systems capture methane emissions by processing manure in oxygen-free environments. These systems convert organic matter into biogas, which can be used as a renewable energy source for electricity, heating, or cooking. The process not only reduces methane emissions but also produces nutrient-rich digestate that can be used as a fertilizer, enhancing soil health without relying on synthetic inputs.

Composting

Composting is another effective method for managing manure. Properly aerated composting systems minimize methane emissions by creating aerobic conditions that support the decomposition of organic matter. The end product is a stable, nutrient-rich material that improves soil structure and fertility, contributing to carbon storage in agricultural soils.

Storage and Application Techniques

Improved storage and application methods also play a role in reducing emissions. Covered storage facilities limit the release of methane and ammonia during manure decomposition. Precision application techniques, such as injecting manure directly into the soil, reduce nitrous oxide emissions by limiting nitrogen volatilization and runoff. These methods also ensure that nutrients are utilized effectively by crops, improving productivity.

Efficient Irrigation Practices

Efficient irrigation practices are essential for reducing nitrous oxide emissions associated with the overuse of water and fertilizers. Nitrous oxide is released when excess nitrogen, applied as fertilizer, undergoes microbial processes in waterlogged soils. Optimizing irrigation systems can mitigate these emissions while conserving water resources.

Drip Irrigation

Drip irrigation delivers water directly to the roots of plants through a network of pipes and emitters, minimizing water waste and soil saturation. This method reduces the potential for nitrogen leaching and denitrification, processes that contribute to nitrous oxide emissions. Drip irrigation also improves water use efficiency, ensuring that crops receive the optimal amount of moisture.

Scheduling and Monitoring

Effective irrigation scheduling, guided by weather forecasts, soil moisture sensors, and crop needs, prevents overwatering and reduces the risk of emissions. Advanced monitoring technologies, such as remote sensing and smart irrigation systems, provide real-time data to optimize water use and minimize environmental impacts.

Fertilizer Management

Efficient irrigation practices are closely linked to fertilizer management. Applying fertilizers in synchrony with irrigation, using techniques like fertigation, ensures that nutrients are delivered directly to the root zone, reducing losses to the atmosphere and waterways. These methods improve nitrogen use efficiency and decrease the risk of nitrous oxide emissions.

Synergies and Co-Benefits

The integration of improved manure management and efficient irrigation practices provides multiple co-benefits beyond emission reductions. For example, anaerobic digestion systems generate renewable energy, contributing to energy security and reducing reliance on fossil fuels. Composting enhances soil organic matter, promoting soil health and productivity.

Efficient irrigation practices conserve water resources, supporting climate adaptation in regions facing water scarcity. These methods also reduce production costs by optimizing resource use, improving the economic sustainability of agricultural systems.

Additionally, these practices align with global sustainability goals, including the United Nations SDGs, by addressing climate action, water management, and sustainable agriculture. Their implementation supports resilient agricultural landscapes that can adapt to changing environmental conditions.

Challenges and Opportunities

Despite their benefits, the adoption of these practices faces several challenges. The upfront costs of anaerobic digestion systems and drip irrigation infrastructure can be prohibitive for smallholder farmers. Limited access to technical knowledge and training also hinders widespread adoption, particularly in developing regions.

Policymakers and stakeholders can address these challenges by providing financial incentives, such as subsidies or low-interest loans, to support the transition to sustainable practices. Capacity-building programs, including farmer training and extension services, are essential for disseminating knowledge and encouraging adoption.

Chapter 4: Forests: Sustainable Management and Enhancement of Carbon Stocks

Forests are critical ecosystems that play an essential role in mitigating climate change through carbon sequestration and storage. This chapter examines the importance of forests in the global carbon cycle, highlighting their capacity to absorb and store large quantities of carbon dioxide. It explores sustainable forest management practices, emphasizing strategies that balance conservation and utilization to maintain ecological and economic value.

The chapter also discusses approaches to enhance forest carbon stocks, such as afforestation, reforestation, and forest restoration, and evaluates the potential of these interventions to contribute to global climate goals. Finally, the chapter includes an analysis of grassland management practices that complement forest ecosystems in carbon sequestration. Together, these strategies illustrate the transformative potential of forests in addressing climate challenges and promoting sustainable development.

Importance of Forests in Carbon Cycling and Storage

Forests are vital components of the global carbon cycle, playing a dual role as significant carbon sinks and sources, depending on how they are managed. Through the process of photosynthesis, forests absorb carbon dioxide (CO_2) from the atmosphere and store it in their biomass, soils, and organic matter, contributing substantially to mitigating climate change. Their role in carbon cycling and storage is critical to maintaining the Earth's climate balance.

Carbon Sequestration in Forests

Forests are among the largest terrestrial carbon sinks, with approximately 45% of terrestrial carbon stored within their ecosystems. This carbon is held in three primary pools: above-ground biomass (trees, shrubs, and understory plants), below-ground biomass (roots), and soil organic matter. Tropical forests, such as the Amazon and Congo basins, store the largest amount of carbon due to their dense vegetation and rapid growth rates. Temperate and boreal forests also contribute significantly to global carbon storage, with boreal forests storing vast amounts of carbon in their soils and peatlands.

The capacity of forests to sequester carbon is dynamic, varying with factors such as forest type, age, and health. Young, growing forests typically absorb carbon at higher rates due to their rapid biomass accumulation, while mature forests store more carbon in their established biomass and soil. Sustainable forest management ensures that forests continue to act as carbon sinks, maintaining their ability to capture and store CO_2 over time.

Forests and the Carbon Cycle

Forests play a pivotal role in regulating the carbon cycle by balancing carbon absorption and release. During photosynthesis, trees and plants take in CO_2, converting it into organic compounds that form the building blocks of their growth. This process not only reduces atmospheric carbon levels but also supports biodiversity and ecosystem functioning.

Forests also influence the carbon cycle through respiration and decomposition. While live trees release small amounts of CO_2 through respiration, the decay of organic matter and deadwood contributes to carbon release. When forests are disturbed—through deforestation, logging, or wildfires—they can become net carbon sources, emitting more carbon than they sequester. Preventing such disturbances is critical for maintaining their role as carbon sinks.

Soil Carbon Storage in Forests

Forest soils are significant reservoirs of carbon, often holding more carbon than the vegetation above them. This carbon is stored in organic matter, such as decaying leaves and roots, and in stable soil compounds. Practices like minimizing soil disturbance and preserving forest floor litter are essential for maintaining soil carbon storage. Degradation of forest soils through unsustainable practices can lead to significant carbon emissions, emphasizing the need for careful soil management.

Global Climate Goals and Forests

Forests are integral to achieving global climate goals, such as those outlined in the Paris Agreement. They offer nature-based solutions for reducing greenhouse gas emissions, enhancing resilience, and supporting biodiversity conservation. Initiatives like afforestation, reforestation, and forest restoration further enhance their carbon sequestration potential, contributing to net-zero targets and sustainable development objectives.

Sustainable Forest Management Practices

Sustainable forest management (SFM) is a critical approach to ensuring that forests continue to provide essential ecological, economic, and social benefits while maintaining their role as carbon sinks. By balancing conservation with utilization, SFM promotes the responsible stewardship of forest resources, aligning short-term needs with long-term sustainability. This practice is integral to ecosystem-based mitigation and achieving global climate and biodiversity goals.

Principles of Sustainable Forest Management

The foundation of SFM lies in managing forests to meet present needs without compromising their ability to serve future generations. It incorporates ecological, social, and economic considerations, ensuring that forest ecosystems remain productive and resilient. Key principles of SFM include maintaining forest cover, preserving biodiversity, enhancing soil and water resources, and promoting equitable resource distribution.

Selective Logging

Selective logging is a sustainable alternative to clear-cutting, where only specific trees are harvested based on criteria such as age, species, or size. This practice minimizes disturbance to the forest ecosystem, allowing remaining trees and understory vegetation to continue carbon sequestration and provide habitat for wildlife. Selective logging also reduces soil erosion and maintains water quality by preserving the forest canopy and root systems.

To ensure its effectiveness, selective logging requires careful planning and monitoring. Logging operations must adhere to sustainable harvesting limits and avoid high-conservation-value areas. Certification schemes like those offered by the Forest Stewardship Council (FSC) encourage adherence to sustainable logging standards.

Agroforestry and Community Forestry

Agroforestry integrates trees into agricultural systems, combining the benefits of forestry and farming. This practice enhances forest conservation by reducing the pressure to clear forests for agriculture. Agroforestry systems provide alternative income sources, such as fruits, nuts, and timber, while sequestering carbon and improving soil health.

Community forestry empowers local communities to manage forest resources sustainably. By involving communities in decision-making and resource management, this approach ensures equitable benefit-sharing and fosters local stewardship. Community-led initiatives often align with traditional knowledge and practices, enhancing the resilience of forest ecosystems.

Reforestation and Afforestation

Reforestation and afforestation are key practices in SFM, aimed at increasing forest cover and enhancing carbon stocks. Reforestation involves planting trees in degraded or deforested areas, restoring ecosystems and sequestering carbon. Afforestation, on the other hand, entails creating forests in areas that were not previously forested, expanding carbon sequestration potential.

Both practices require careful species selection to ensure ecological compatibility and resilience. Native species are often prioritized to promote biodiversity and adapt to local environmental conditions. Monitoring and maintenance are critical to the success of reforestation and afforestation projects, as young forests are vulnerable to pests, diseases, and climate stressors.

Forest Fire Management

Forest fires pose significant threats to ecosystems and carbon storage. SFM incorporates proactive measures to prevent and manage wildfires, such as controlled burns, firebreaks, and early detection systems. Controlled burns mimic natural fire cycles, reducing the risk of catastrophic wildfires while promoting ecological processes that depend on periodic fire.

Community involvement in fire management is crucial, as local knowledge often provides valuable insights into fire behavior and prevention. Training programs and resources for communities can enhance their capacity to respond to fire threats, reducing the likelihood of large-scale forest loss.

Certification and Monitoring

SFM practices are often guided by certification standards that promote responsible forest management. Programs like the FSC and the Programme for the Endorsement of Forest Certification (PEFC) provide frameworks for sustainable harvesting, biodiversity conservation, and social equity. These certifications ensure accountability and encourage market incentives for sustainably sourced products.

Monitoring and evaluation are integral to SFM, as they enable adaptive management based on environmental and social feedback. Technologies like remote sensing and Geographic Information Systems (GIS) enhance monitoring capabilities, providing real-time data on forest health, cover, and carbon stocks.

Enhancing Forest Carbon Stocks Through Afforestation, Reforestation, and Forest Restoration

Forests play a vital role in carbon sequestration, acting as natural carbon sinks that absorb and store carbon dioxide from the atmosphere. Enhancing forest carbon stocks through afforestation, reforestation, and forest restoration is a crucial strategy for mitigating climate change. These practices not only increase carbon storage but also provide co-benefits such as biodiversity conservation, improved ecosystem services, and resilience to environmental changes.

Afforestation

Afforestation involves planting trees in areas that were not previously forested. This practice is particularly valuable in expanding forest cover and establishing new carbon sinks in regions where land has been degraded or underutilized. By transforming barren landscapes into thriving ecosystems, afforestation sequesters significant amounts of carbon in tree biomass, roots, and soil organic matter.

Selecting appropriate tree species is essential for the success of afforestation projects. Native species are often preferred, as they are better adapted to local environmental conditions and support biodiversity. Additionally, mixed-species plantations provide greater ecological benefits and resilience compared to monocultures. Careful planning and management are required to address potential challenges such as water competition, soil degradation, and land-use conflicts.

Afforestation also contributes to climate adaptation by reducing soil erosion, improving water infiltration, and moderating local climates. These benefits make it a valuable component of integrated land management strategies.

Reforestation

Reforestation involves replanting trees in areas where forests have been previously cleared or degraded. This practice restores the ecological functions of forests, enabling them to regain their role as carbon sinks. Reforestation projects often target deforested lands, abandoned agricultural fields, and areas affected by logging or natural disasters.

Successful reforestation requires careful site selection and preparation. Degraded soils may need enrichment with organic matter or nutrients to support tree growth. Planting diverse tree species can enhance biodiversity, improve ecosystem stability, and increase carbon sequestration potential. Fast-growing species are often included to achieve rapid carbon gains, while slow-growing species contribute to long-term storage.

Community involvement is critical for reforestation projects, ensuring that local needs and knowledge are integrated into planning and implementation. For example, community-led initiatives that combine reforestation with agroforestry or sustainable harvesting practices provide dual benefits for ecosystems and livelihoods.

Forest Restoration

Forest restoration focuses on recovering the ecological integrity and functionality of degraded forests. Unlike afforestation and reforestation, which involve planting new trees, restoration often emphasizes natural regeneration. This approach allows forests to recover through the growth of existing vegetation, seed dispersal, and natural succession.

Restoration practices include controlling invasive species, reducing human disturbances, and protecting regenerating areas from grazing or logging. In some cases, assisted regeneration may be necessary, involving selective planting or the introduction of native species to accelerate recovery.

Restored forests provide a range of benefits beyond carbon sequestration. They improve water quality, enhance wildlife habitats, and stabilize local climates. Restoring degraded forest landscapes also reduces the risk of desertification and supports ecosystem resilience in the face of climate change.

Monitoring and Maintenance

The success of afforestation, reforestation, and restoration efforts depends on ongoing monitoring and maintenance. Regular assessments of tree survival rates, growth, and carbon sequestration performance are essential to measure the effectiveness of these projects. Remote sensing technologies and GIS provide valuable tools for tracking forest cover changes and estimating carbon storage.

Maintenance activities, such as thinning, pruning, and controlling pests or diseases, are necessary to ensure healthy forest growth. Addressing challenges such as water scarcity, land-use pressures, and community conflicts is also critical for sustaining these efforts.

Challenges and Opportunities

While enhancing forest carbon stocks offers significant benefits, it is not without challenges. Land availability, funding limitations, and competing land uses can hinder large-scale implementation. Additionally, poorly planned projects, such as afforestation on unsuitable land, may result in ecological harm or limited carbon gains.

To overcome these challenges, governments, organizations, and stakeholders must collaborate to develop effective policies and financial mechanisms. Incentives like carbon credits, payments for ecosystem services, and grants for restoration projects can encourage adoption and ensure long-term sustainability. Engaging local communities and incorporating traditional knowledge into project design are also key to achieving success.

Grassland Management Practices That Increase Carbon Sequestration

Grasslands are extensive ecosystems with significant potential to sequester carbon in soils and vegetation. Proper management of grasslands can enhance their role as carbon sinks, contributing to climate mitigation while supporting biodiversity and livelihoods. Effective grassland management practices focus on maintaining vegetation cover, improving soil health, and promoting sustainable land use.

Importance of Grasslands in Carbon Sequestration

Grasslands cover approximately 40% of the Earth's terrestrial area and store a substantial amount of carbon in their soils. Unlike forests, where carbon is stored primarily in above-ground biomass, grasslands sequester most of their carbon below ground in root systems and soil organic matter. This makes grasslands a stable and resilient carbon sink, as soil carbon is less susceptible to disturbances such as fires or harvesting.

However, overgrazing, land conversion, and unsustainable agricultural practices can degrade grasslands, reducing their carbon storage capacity and releasing stored carbon into the atmosphere. Implementing sustainable grassland management practices is essential for preserving and enhancing their carbon sequestration potential.

Sustainable Grazing Practices

Grazing is a common land use in grasslands, but improper grazing management can lead to soil erosion, compaction, and carbon loss. Sustainable grazing practices ensure that livestock grazing does not exceed the land's carrying capacity, maintaining vegetation cover and soil health.

Rotational Grazing

Rotational grazing involves dividing grasslands into smaller paddocks and rotating livestock between them. This approach allows grazed areas to recover, promoting plant regrowth and maintaining soil organic matter. Rotational grazing also prevents overgrazing and soil compaction, which can degrade carbon storage capacity.

Adaptive Grazing

Adaptive grazing is a flexible approach that adjusts grazing intensity and timing based on environmental conditions, such as rainfall and vegetation growth. This method optimizes forage use while preserving soil and plant health, contributing to long-term carbon sequestration.

Restoration of Degraded Grasslands

Restoring degraded grasslands is a critical strategy for increasing their carbon storage capacity. Restoration practices focus on re-establishing vegetation, improving soil health, and addressing the causes of degradation.

Reseeding Native Species

Introducing native grasses and plants through reseeding enhances biodiversity and restores natural ecosystem functions. Native species are often better adapted to local conditions, promoting resilience and sustainable carbon sequestration.

Erosion Control

Erosion control measures, such as planting cover crops or using physical barriers, stabilize soil and prevent carbon loss. These practices protect existing soil organic matter and provide conditions for increased carbon storage.

Integrated Crop-Livestock Systems

Combining crop and livestock systems on grasslands can enhance carbon sequestration by diversifying land use and improving resource efficiency. For example, integrating forage crops with livestock grazing provides continuous ground cover, reduces soil erosion, and increases soil organic matter. This approach also enhances nutrient cycling, as livestock manure contributes organic matter and nutrients to the soil.

Conservation of Soil Organic Matter

Maintaining and enhancing soil organic matter is crucial for carbon sequestration in grasslands. Practices such as minimal soil disturbance, cover cropping, and organic amendments improve soil structure and carbon storage.

No-Till or Reduced-Till Practices

Minimizing soil disturbance through no-till or reduced-till practices prevents the release of stored carbon and enhances soil organic matter accumulation. These methods also improve water retention and reduce erosion, creating favorable conditions for carbon sequestration.

Organic Amendments

Applying organic materials, such as compost or manure, to grasslands enriches soil organic matter and promotes microbial activity. These amendments increase soil carbon storage and support plant growth.

Protection and Conservation

Protecting existing grasslands from degradation and conversion is critical for preserving their carbon sequestration capacity. Legal and policy measures, such as establishing protected areas or implementing land-use regulations, ensure that grasslands remain intact and continue to function as carbon sinks.

Co-Benefits of Grassland Management

Sustainable grassland management offers numerous co-benefits beyond carbon sequestration. Healthy grasslands support biodiversity by providing habitats for wildlife and pollinators. They also contribute to water regulation, preventing floods and maintaining groundwater recharge. Additionally, sustainable practices improve forage quality and productivity, enhancing the livelihoods of livestock farmers.

Challenges and Opportunities

While grassland management has immense potential, its implementation faces challenges, such as limited financial resources, technical knowledge gaps, and competing land-use demands. Addressing these barriers requires collaborative efforts among governments, landowners, and communities.

Incentives such as PES and carbon credit schemes can encourage adoption of sustainable practices. Capacity-building programs and research initiatives are also critical for developing innovative management techniques and disseminating knowledge to land managers.

Chapter 5: Peatlands: Restoring and Conserving Carbon-Rich Wetlands

Peatlands are among the most carbon-rich ecosystems on Earth, storing vast amounts of carbon in their waterlogged soils and playing a crucial role in regulating the global climate. Despite covering only 3% of the Earth's land surface, peatlands store more carbon than all the world's forests combined, making their protection and restoration essential for effective climate mitigation.

This chapter explores the unparalleled carbon storage potential of peatlands and the threats posed by drainage, land conversion, and degradation. It examines strategies for restoring and conserving these ecosystems, emphasizing the importance of rewetting degraded peatlands, sustainable management practices, and global efforts to protect these vital landscapes. By showcasing the ecological and socio-economic benefits of peatland conservation, this chapter highlights the pivotal role of peatlands in achieving climate goals and promoting ecosystem resilience.

Carbon Storage Potential of Peatlands

Peatlands are one of the most significant natural carbon sinks on Earth, playing a crucial role in mitigating climate change. These wetlands, characterized by waterlogged soils rich in partially decomposed organic matter, store vast amounts of carbon that have accumulated over thousands of years. Despite covering only 3% of the Earth's land surface, peatlands hold more carbon than all the world's forests combined, making them a vital component of the global carbon cycle.

How Peatlands Store Carbon

Peatlands accumulate carbon through the slow decomposition of plant material under waterlogged and anaerobic conditions. This process creates peat, a dense, carbon-rich soil that stores organic carbon for millennia. The water saturation in peatlands inhibits microbial activity, which slows down the breakdown of organic matter, allowing carbon to build up over time. On average, peat accumulates at a rate of 1 millimeter per year, yet it can store carbon on a scale that far surpasses other ecosystems due to its long-term stability.

Carbon Storage in Tropical and Boreal Peatlands

Peatlands exist in various climates, with tropical and boreal peatlands being the most prominent. Tropical peatlands, found in regions like Southeast Asia, the Amazon, and the Congo Basin, store enormous quantities of carbon in thick deposits of peat. For example, the peatlands of Indonesia and Malaysia are estimated to hold more than 100 gigatons of carbon.

Boreal and temperate peatlands, located in regions such as Canada, Russia, and northern Europe, also store significant amounts of carbon, primarily in their soils. These peatlands have a slower rate of carbon accumulation due to cooler temperatures, but they cover larger areas, contributing substantially to global carbon storage. Combined, these peatland types account for over 600 gigatons of stored carbon, equivalent to nearly twice the amount of carbon stored in the world's forests.

Threats to Peatland Carbon Storage

Despite their importance, peatlands are under significant threat from human activities, including drainage, agriculture, forestry, and peat extraction. When peatlands are drained for agriculture or other land uses, the water table lowers, exposing peat to oxygen. This exposure accelerates decomposition, releasing stored carbon as CO_2 and, in some cases, CH_4. Drained and degraded peatlands currently contribute about 5% of global anthropogenic CO_2 emissions annually.

Fires, often used to clear peatlands for agriculture, further exacerbate carbon loss. Peat fires can smolder for weeks, releasing large amounts of greenhouse gases and particulate matter, which have severe environmental and health impacts.

Importance of Conserving Intact Peatland

Preserving intact peatlands is one of the most effective ways to maintain their carbon storage capacity. Undisturbed peatlands continue to act as carbon sinks, slowly accumulating carbon over time while providing additional ecosystem services such as water regulation, biodiversity conservation, and climate resilience. By preventing drainage and degradation, intact peatlands can contribute to achieving global climate goals.

Restoration of Degraded Peatlands

Restoring degraded peatlands is a critical strategy for mitigating carbon emissions and enhancing their role as carbon sinks. Rewetting, the process of raising the water table to restore anaerobic conditions, is a key restoration technique. Rewetting halts the decomposition of peat and reduces CO_2 emissions, allowing peatlands to transition from a carbon source to a sink over time. Additionally, re-establishing native vegetation, such as sphagnum mosses, promotes the accumulation of new peat and further enhances carbon sequestration.

Global Significance of Peatlands

Peatlands are integral to global climate strategies due to their unparalleled carbon storage potential. Their conservation and restoration are emphasized in international frameworks such as the Paris Agreement and the Ramsar Convention on Wetlands. Recognizing the economic and ecological value of peatlands, governments and organizations worldwide are increasingly investing in peatland protection and rehabilitation.

Threats to Peatlands and Their Contribution to Greenhouse Gas Emissions

Peatlands are among the most effective natural carbon sinks, but they are increasingly under threat from human activities and climate change. These threats not only compromise the ecological integrity of peatlands but also lead to significant GHG emissions. When disturbed, peatlands, which store vast amounts of carbon, transition from carbon sinks to carbon sources, contributing to global climate change.

Drainage and Agricultural Conversion

One of the most pervasive threats to peatlands is drainage for agricultural and forestry purposes. Lowering the water table through drainage exposes peat to oxygen, accelerating the decomposition of organic matter. This process releases stored carbon as CO_2 and, in some cases, CH_4. Drained peatlands are estimated to contribute approximately 5% of global anthropogenic CO_2 emissions annually, despite covering only 0.4% of the Earth's land area.

Agricultural conversion often involves clearing peatlands to grow crops like oil palm, soybeans, or rice. These activities not only release GHGs but also degrade soil quality, reducing the long-term viability of the land for farming. The drainage and tilling of peatlands amplify carbon emissions while increasing the risk of subsidence, flooding, and further degradation.

Peat Extraction

Peat is extracted for various purposes, including horticulture, fuel, and industrial use. This extraction removes layers of peat that have accumulated over thousands of years, releasing large quantities of CO_2 in the process. The harvesting of peat for fuel is particularly carbon-intensive, as it involves burning peat directly, which produces emissions comparable to fossil fuels.

In many regions, peat extraction is poorly regulated, leading to widespread degradation of peatlands. The long-term consequences include a loss of biodiversity, reduced ecosystem services, and increased GHG emissions, as the exposed peat continues to oxidize after extraction.

Fires in Peatlands

Fires are another major threat to peatlands, particularly in tropical regions. Fires are often deliberately set to clear land for agriculture or plantation development. Once ignited, peat fires can smolder for weeks or months, burning deep into the peat layers and releasing massive amounts of CO_2, CH_4, and particulate matter. These emissions significantly contribute to air pollution and climate change.

Peatland fires are exacerbated by drainage, which dries out the peat and makes it highly flammable. Regions such as Southeast Asia have experienced severe peatland fire events, resulting in transboundary haze pollution, health crises, and economic losses.

Climate Change Impacts

Climate change poses an additional threat to peatlands by altering hydrological and temperature regimes. Rising temperatures can increase evaporation rates, drying out peatlands and making them more susceptible to decomposition and fires. Prolonged droughts, a consequence of climate change, further exacerbate the drying of peatlands, reducing their ability to act as carbon sinks.

Melting permafrost in northern peatlands is another emerging concern. As permafrost thaws, previously frozen organic material becomes exposed to microbial activity, releasing CO_2 and CH_4 into the atmosphere. This feedback loop amplifies global warming and undermines the stability of these critical ecosystems.

Urbanization and Infrastructure Development

Urban expansion and infrastructure projects, such as roads, pipelines, and industrial facilities, encroach upon peatlands, leading to habitat loss and fragmentation. These developments often involve drainage and excavation, resulting in significant carbon emissions. Urbanization also increases the risk of pollution from chemicals and waste, further degrading peatland ecosystems.

Cumulative Impact on Greenhouse Gas Emissions

The combined effects of these threats result in a substantial contribution to global GHG emissions. Drained and degraded peatlands emit billions of tons of CO_2 annually, with tropical peatlands being particularly significant sources due to their high carbon density. Fires and land-use changes exacerbate these emissions, contributing to a cycle of degradation and climate impacts.

Strategies for Peatland Restoration and Conservation

Peatlands are vital ecosystems that store vast amounts of carbon and provide essential ecological services. However, the degradation of peatlands through drainage, agricultural conversion, and peat extraction has transformed many of these ecosystems from carbon sinks into significant sources of GHG emissions. Effective strategies for peatland restoration and conservation are crucial to reversing this trend, mitigating climate change, and preserving biodiversity and ecosystem resilience.

Rewetting Degraded Peatlands

Rewetting is a foundational strategy for restoring degraded peatlands. By raising the water table to restore waterlogged conditions, rewetting prevents further oxidation of peat, halting the release of stored carbon as CO_2. Restoring anaerobic conditions also creates an environment that supports the regrowth of peat-forming vegetation, such as sphagnum mosses, which can resume carbon accumulation over time.

Rewetting often involves blocking drainage channels with dams, bunds, or water retention structures. These measures reduce water outflow and restore the natural hydrology of peatlands. In areas where natural water sources are unavailable, artificial rehydration using pumped water may be employed.

Revegetation and Assisted Regeneration

Revegetation is essential for accelerating peatland recovery. Planting native, peat-forming vegetation, such as sphagnum mosses in boreal and temperate peatlands or sedges and wetland trees in tropical regions, helps stabilize the soil and promote peat formation. These plants also enhance biodiversity and provide habitats for a wide range of species.

Assisted regeneration involves actively supporting natural recovery processes. This can include removing invasive species that compete with native vegetation, protecting regenerating areas from grazing or logging, and creating conditions conducive to natural seed dispersal. In some cases, seeding or planting nursery-grown seedlings may be necessary to re-establish vegetation.

Fire Prevention and Management

Peatland fires are a significant threat to both restored and intact peatlands, releasing large amounts of GHGs and causing widespread environmental damage. Fire prevention strategies include maintaining high water levels to reduce the flammability of peat and establishing buffer zones with fire-resistant vegetation.

Community involvement is critical for effective fire management. Local communities can be engaged in monitoring and reporting fire risks, conducting controlled burns under safe conditions, and implementing traditional fire prevention practices. Early detection systems using remote sensing and satellite imagery are valuable tools for identifying and addressing fire outbreaks before they spread.

Sustainable Land Use Practices

Conservation strategies must address the root causes of peatland degradation, including unsustainable land use. Shifting from drainage-based agriculture to wet agriculture, or paludiculture, allows farming in rewetted peatlands. Paludiculture involves cultivating wetland-adapted crops, such as reeds, cattails, and sphagnum moss, which can be harvested for fiber, bioenergy, or horticulture without damaging the peat.

Agroforestry systems, which integrate trees with agricultural practices, are another option for sustainable land use in peatland areas. These systems balance productivity with conservation by reducing soil disturbance, enhancing biodiversity, and maintaining carbon storage.

Policy and Governance

Effective policies and governance frameworks are essential for peatland restoration and conservation. Governments play a central role in establishing regulations that protect peatlands, such as restricting drainage, banning peat extraction, and designating peatlands as protected areas.

Economic incentives, such as PES and carbon credit schemes, encourage landowners and communities to adopt sustainable practices. International frameworks like the Ramsar Convention on Wetlands and the Paris Agreement provide global support for peatland conservation, fostering collaboration and resource mobilization across borders.

Community Engagement and Education

Engaging local communities is a cornerstone of successful peatland restoration. Many communities depend on peatlands for their livelihoods and must be involved in the planning and implementation of conservation projects. Providing education and training on sustainable practices, fire prevention, and alternative income-generating activities ensures that restoration efforts are socially inclusive and economically viable.

Participatory approaches, such as community-led monitoring and management, enhance local ownership and accountability. These initiatives empower communities to become active stewards of peatlands, aligning conservation goals with their socio-economic needs.

Monitoring and Evaluation

Ongoing monitoring and evaluation are critical for assessing the success of peatland restoration efforts. Remote sensing technologies, such as satellite imagery and drones, enable the tracking of vegetation growth, water levels, and carbon fluxes. Ground-based measurements, including soil sampling and biodiversity surveys, complement these tools by providing detailed data on ecosystem health.

Adaptive management, which involves refining strategies based on monitoring results, ensures that restoration efforts remain effective in the face of changing environmental conditions. Transparency in reporting and data sharing also fosters collaboration among stakeholders and enhances accountability.

Global Efforts to Protect and Restore Peatlands

Peatlands are critical ecosystems that store vast amounts of carbon, support biodiversity, and provide essential ecosystem services. Recognizing their global importance, efforts to protect and restore peatlands have gained momentum over recent decades. International agreements, national initiatives, and collaborative projects are driving conservation and restoration efforts to mitigate climate change and ensure sustainable land use.

International Frameworks and Agreements

Several international frameworks underscore the importance of peatlands in addressing climate change and biodiversity loss. The Ramsar Convention on Wetlands, established in 1971, is one of the earliest global agreements promoting the conservation and wise use of wetlands, including peatlands. Countries that are signatories commit to protecting and restoring peatlands of international significance.

The Paris Agreement further highlights peatlands' role in climate mitigation by encouraging countries to include wetland restoration in their nationally determined contributions (NDCs). Protecting peatlands aligns with the agreement's goals to reduce GHG emissions and enhance ecosystem resilience.

The United Nations SDGs also integrate peatland conservation into global sustainability targets. Specifically, goals related to climate action (SDG 13), life on land (SDG 15), and clean water (SDG 6) emphasize the importance of maintaining healthy peatland ecosystems.

National Initiatives

Many countries with significant peatland coverage have launched national initiatives to protect and restore these ecosystems. For example, Indonesia, which holds one of the largest tropical peatland reserves, established the Peatland Restoration Agency (BRG) in 2016. This agency aims to restore over two million hectares of degraded peatlands and reduce the frequency of peat fires that contribute to regional haze and global GHG emissions.

In Europe, countries such as Finland, Germany, and the United Kingdom have implemented large-scale peatland restoration programs. These efforts often include rewetting degraded peatlands, reintroducing native vegetation, and integrating peatland conservation into agricultural and forestry policies. Scotland, for instance, has committed to restoring 250,000 hectares of peatlands by 2030 as part of its climate action plan.

Collaborative Projects and Research

International collaborations and research initiatives are playing a pivotal role in advancing peatland restoration. The Global Peatlands Initiative (GPI), launched by the United Nations Environment Programme (UNEP), brings together governments, non-governmental organizations (NGOs), and research institutions to protect peatlands worldwide. The GPI focuses on raising awareness, sharing best practices, and mobilizing resources for peatland conservation.

The International Mire Conservation Group (IMCG) and the Society for Ecological Restoration (SER) are also instrumental in promoting research and capacity building. These organizations facilitate knowledge exchange and provide technical guidance for restoring degraded peatlands across diverse regions.

Funding Mechanisms and Economic Incentives

Securing funding is essential for the success of peatland protection efforts. International funding mechanisms, such as the GCF and the GEF, provide financial support for large-scale restoration projects. These funds help countries implement peatland conservation initiatives, particularly in developing regions where resources are limited.

Economic incentives, such as PES and carbon credit schemes, encourage landowners and communities to adopt sustainable peatland management practices. For example, carbon markets enable countries and organizations to offset emissions by investing in peatland restoration, creating economic opportunities while reducing GHG emissions.

Challenges and Opportunities

Despite growing efforts, significant challenges remain in protecting and restoring peatlands globally. Land-use conflicts, limited technical expertise, and insufficient funding often hinder progress. Addressing these barriers requires stronger international cooperation, improved governance, and increased public and private investment.

Opportunities for scaling up peatland conservation include leveraging advancements in remote sensing and GIS to monitor restoration progress. Additionally, integrating peatland conservation into broader climate and biodiversity strategies enhances its visibility and funding potential.

Chapter 6: Marine and Coastal Ecosystems: Blue Carbon Solutions

Marine and coastal ecosystems, including mangroves, seagrass meadows, and tidal salt marshes, play a vital role in mitigating climate change through their ability to sequester and store atmospheric carbon, commonly referred to as "blue carbon." These ecosystems are among the most efficient natural carbon sinks, capturing carbon at rates far exceeding those of terrestrial forests.

This chapter explores the importance of marine and coastal ecosystems in carbon sequestration and highlights their ecological, economic, and social value. It examines strategies for the conservation and restoration of blue carbon ecosystems, emphasizing the role of Marine Protected Areas (MPAs) in preserving their integrity. Additionally, it addresses the challenges of protecting these ecosystems, including threats from human activities, climate change, and coastal development. By showcasing the potential of blue carbon solutions, this chapter underscores their critical role in global climate strategies and sustainable coastal management.

Importance of Coastal and Marine Ecosystems in Carbon Sequestration

Coastal and marine ecosystems, often referred to as blue carbon ecosystems, are among the most effective natural systems for carbon sequestration. These ecosystems include mangroves, seagrass meadows, tidal salt marshes, and kelp forests, all of which capture and store atmospheric CO_2 in their biomass and underlying sediments. Despite occupying only a small fraction of the Earth's surface, these ecosystems play an outsized role in mitigating climate change and maintaining global carbon balance.

High Carbon Sequestration Efficiency

Blue carbon ecosystems sequester carbon at rates significantly higher than terrestrial forests. Mangroves, for instance, can store up to four times more carbon per hectare than tropical rainforests. Seagrass meadows and tidal salt marshes similarly have high carbon burial rates, as their sediments accumulate organic material over centuries or even millennia. The ability of these ecosystems to store carbon in both above-ground biomass (leaves, stems, and roots) and sediments enhances their sequestration potential.

Sequestered carbon in these ecosystems is often referred to as "blue carbon." Unlike terrestrial carbon, which is stored mainly in trees and plants, blue carbon is predominantly stored in sediments, where anaerobic conditions slow decomposition and preserve organic carbon for long periods.

Global Carbon Storage Contribution

Despite their limited geographic extent, blue carbon ecosystems contribute significantly to global carbon storage. Mangroves, covering only 0.1% of the Earth's surface, account for roughly 10-15% of the carbon stored in coastal sediments. Seagrass meadows, which span about 300,000 square kilometers globally, store an estimated 10% of the organic carbon buried in the ocean each year. These figures underscore the disproportionate importance of coastal and marine ecosystems in global carbon budgets.

Climate Mitigation and Adaptation

In addition to their role in carbon sequestration, coastal and marine ecosystems provide essential services that contribute to climate adaptation. Mangroves, for example, act as natural buffers against storm surges, protecting coastal communities from flooding and erosion. Seagrass meadows stabilize sediments and improve water quality by filtering pollutants, while tidal salt marshes reduce the impacts of wave energy on shorelines. These ecosystem services are critical for building resilience to climate-related challenges such as rising sea levels and extreme weather events.

Threats to Blue Carbon Ecosystems

Despite their importance, coastal and marine ecosystems are under significant threat from human activities. Coastal development, pollution, overfishing, and aquaculture lead to habitat degradation and loss, reducing their carbon storage capacity. Additionally, climate change exacerbates these threats through rising sea levels, ocean acidification, and increased water temperatures, which can disrupt the delicate balance of these ecosystems.

When degraded or destroyed, blue carbon ecosystems release stored carbon back into the atmosphere, contributing to greenhouse gas emissions. For example, the conversion of mangroves to aquaculture ponds can release decades' worth of sequestered carbon in a matter of years, turning these ecosystems into net carbon sources.

Conservation and Restoration Efforts

Protecting and restoring coastal and marine ecosystems is essential to maintaining their carbon sequestration functions. Conservation strategies include establishing MPAs, regulating coastal development, and implementing sustainable fishing practices. Restoration projects, such as replanting mangroves or restoring seagrass meadows, have demonstrated success in enhancing carbon storage capacity while delivering additional ecological and socio-economic benefits.

Conservation and Restoration of Mangroves, Tidal Salt Marshes, and Seagrass Beds

Mangroves, tidal salt marshes, and seagrass beds are vital components of coastal and marine ecosystems, renowned for their unparalleled carbon sequestration capabilities and ecosystem services. These blue carbon ecosystems play a crucial role in mitigating climate change, supporting biodiversity, and protecting coastal communities. However, human activities and climate change threaten their existence, making conservation and restoration efforts essential for sustaining their ecological and socio-economic functions.

Mangroves: Conservation and Restoration

Mangroves, found along tropical and subtropical coastlines, are unique intertidal forests that provide critical carbon storage in both their biomass and sediments. These ecosystems also act as natural buffers against storm surges, reduce coastal erosion, and support diverse marine and terrestrial species.

Conservation Efforts

Mangrove conservation involves protecting existing mangrove forests from deforestation and degradation. Establishing MPAs is a widely adopted strategy to safeguard mangroves, preventing activities like logging, aquaculture, and unsustainable tourism. Community-based conservation initiatives further enhance protection by involving local stakeholders in monitoring and sustainable management practices.

Restoration Techniques

Restoration of mangroves focuses on replanting degraded areas and creating conditions conducive to natural regeneration. Techniques include:

• **Hydrological Restoration**: Restoring natural tidal flows to areas where mangroves have been degraded due to altered water regimes.

• **Direct Planting**: Planting mangrove seedlings or propagules, often involving local communities in planting efforts to foster stewardship.

• **Assisted Natural Regeneration**: Protecting degraded sites to allow natural seed dispersal and regeneration.

Success depends on selecting suitable mangrove species, considering site-specific conditions like salinity and hydrology, and engaging local communities to ensure long-term management and benefits.

Tidal Salt Marshes: Conservation and Restoration

Tidal salt marshes, located in temperate regions, are coastal wetlands dominated by salt-tolerant grasses and shrubs. These ecosystems are vital for carbon sequestration, water filtration, and providing habitats for fish and bird species. They also act as natural flood barriers, protecting inland areas from storm surges and sea-level rise.

Conservation Efforts

The conservation of tidal salt marshes involves limiting land-use changes that lead to habitat loss, such as conversion to agriculture or urban development. Regulatory frameworks and coastal zoning laws can protect these ecosystems from encroachment. Additionally, managing invasive species that outcompete native vegetation is critical to maintaining the ecological balance of salt marshes.

Restoration Techniques

Restoring tidal salt marshes often requires re-establishing natural tidal flows and replanting native vegetation. Key approaches include:

• **Reconnection of Tidal Flow**: Removing barriers such as dikes or levees to restore the natural exchange of seawater and sediment.

• **Vegetation Planting**: Planting native marsh grasses and shrubs to stabilize sediments and promote carbon storage.

• **Sediment Addition**: Adding sediment to subsided areas to raise the marsh surface to a level suitable for vegetation growth.

Monitoring and adaptive management are essential for ensuring the success of restoration projects, particularly in the face of rising sea levels.

Seagrass Beds: Conservation and Restoration

Seagrass beds, submerged meadows of flowering plants found in shallow marine waters, are vital for carbon sequestration, supporting fisheries, and improving water quality. Seagrasses trap carbon in their root systems and sediments, where it remains buried for centuries. They also stabilize the seabed, reduce wave energy, and provide habitats for diverse marine species.

Conservation Efforts

Seagrass conservation involves mitigating threats such as coastal development, pollution, and destructive fishing practices. Establishing MPAs and regulating activities like anchoring and trawling help protect seagrass habitats from physical damage. Public awareness campaigns and stakeholder engagement are crucial for reducing pollution and other human impacts.

Restoration Techniques

Restoration of seagrass beds is challenging but achievable with advanced techniques and careful planning. Methods include:

• **Seed Planting**: Collecting and planting seagrass seeds in suitable areas, ensuring adequate light and sediment stability for growth.

• **Transplantation**: Transplanting healthy seagrass from donor sites to degraded areas, often using frames or mats to anchor the plants.

• **Sediment Stabilization**: Using biodegradable mats or natural materials to stabilize sediments in areas prone to erosion, allowing seagrass to establish.

Restoration projects require monitoring to assess success and address challenges such as grazing by herbivores or water quality issues.

Challenges in Conservation and Restoration

Efforts to conserve and restore mangroves, tidal salt marshes, and seagrass beds face several challenges:

• **Climate Change**: Rising sea levels, warming oceans, and changing salinity levels impact the growth and resilience of these ecosystems.

• **Human Activities**: Coastal development, pollution, and unsustainable resource extraction continue to degrade these habitats.

• **Funding and Resources**: Restoration projects are often resource-intensive, requiring long-term financial and technical support.

Global Collaboration and Opportunities

Collaborative international initiatives, such as the Blue Carbon Initiative and the Global Mangrove Alliance, promote the conservation and restoration of these ecosystems. These efforts involve governments, non-governmental organizations, and local communities, fostering knowledge exchange and funding support. Carbon credit schemes and payments for ecosystem services provide additional economic incentives for conservation.

Role of MPAs in Preserving Blue Carbon Ecosystems

MPAs are designated zones in coastal and marine environments where human activities are regulated to conserve biodiversity, protect habitats, and sustain ecosystem services. MPAs play a crucial role in preserving blue carbon ecosystems, such as mangroves, seagrass meadows, and tidal salt marshes, which are essential for carbon sequestration and climate mitigation. By safeguarding these ecosystems, MPAs contribute to global efforts to combat climate change and support the resilience of marine and coastal communities.

Preserving Blue Carbon Ecosystems

Blue carbon ecosystems are highly efficient at capturing and storing carbon in their biomass and sediments. Mangroves, for instance, sequester large amounts of carbon, storing it both above ground in their roots, trunks, and leaves, and below ground in waterlogged sediments. Similarly, seagrass meadows and tidal salt marshes trap organic carbon in their dense root systems and underlying soils, where it remains for centuries. MPAs are instrumental in protecting these ecosystems from degradation, which can otherwise release stored carbon back into the atmosphere as CO_2.

By reducing anthropogenic pressures such as overfishing, habitat destruction, and pollution, MPAs ensure that blue carbon ecosystems remain intact and continue to function as carbon sinks. These protected areas also provide a buffer against climate change impacts, such as rising sea levels and extreme weather events, which threaten the stability of these ecosystems.

Regulating Harmful Activities

One of the primary functions of MPAs is to regulate activities that pose risks to blue carbon ecosystems. For example, mangrove forests are often cleared for aquaculture, agriculture, or urban development, while seagrass meadows are damaged by anchoring, trawling, and coastal infrastructure projects. Tidal salt marshes face threats from drainage, land reclamation, and pollution.

MPAs establish boundaries and enforce restrictions to prevent such destructive activities. Within these protected zones, regulations may include bans on logging, fishing gear restrictions, and limits on coastal development. These measures reduce physical damage to habitats, maintain water quality, and ensure that blue carbon ecosystems can continue to store carbon effectively.

Enhancing Ecosystem Resilience

MPAs also enhance the resilience of blue carbon ecosystems to climate change. By minimizing human disturbances, MPAs allow ecosystems to recover from past degradation and adapt to changing environmental conditions. For instance, intact mangroves and salt marshes can migrate landward in response to rising sea levels, provided that adjacent land is available and protected. Similarly, seagrass meadows in MPAs are better able to withstand temperature fluctuations and ocean acidification due to reduced stressors like pollution and mechanical damage.

Additionally, MPAs often support research and monitoring programs that provide valuable insights into ecosystem dynamics and climate adaptation strategies. These programs inform the management of blue carbon ecosystems within and beyond protected areas, ensuring their long-term viability.

Supporting Biodiversity and Ecosystem Services

Blue carbon ecosystems are biodiversity hotspots that provide critical habitats for fish, birds, and other marine and terrestrial species. MPAs protect these habitats, enabling species to thrive and maintain healthy populations. This biodiversity, in turn, supports the ecological functions of blue carbon ecosystems, such as nutrient cycling, sediment stabilization, and water filtration.

The ecosystem services provided by blue carbon ecosystems extend beyond carbon sequestration. For example, mangroves reduce the impacts of storm surges, tidal salt marshes act as natural flood defenses, and seagrass meadows improve water quality by trapping sediments and pollutants. MPAs ensure that these services continue to benefit coastal communities, enhancing their resilience to climate-related challenges.

Challenges in MPA Implementation

While MPAs are effective tools for conserving blue carbon ecosystems, their success depends on proper planning, management, and enforcement. Challenges include:

• **Funding and Resources**: Establishing and maintaining MPAs require significant financial and technical resources, which are often limited in developing regions.

• **Community Engagement**: Ensuring local community support is essential for the success of MPAs, as these communities may depend on the protected areas for their livelihoods.

• **Enforcement**: Weak enforcement of regulations can undermine the effectiveness of MPAs, allowing illegal activities to continue.

Addressing these challenges involves securing sustainable funding, fostering stakeholder collaboration, and building local capacity for effective management and enforcement.

Global Initiatives and Collaboration

International efforts, such as the Convention on Biological Diversity (CBD) and the United Nations SDGs, emphasize the importance of MPAs in achieving conservation and climate goals. Collaborative initiatives like the Blue Carbon Initiative and the Global Mangrove Alliance support the establishment and management of MPAs, promoting knowledge sharing and resource mobilization.

Challenges in Protecting Marine and Coastal Ecosystems

Marine and coastal ecosystems, including mangroves, seagrass meadows, and coral reefs, play a critical role in climate regulation, biodiversity conservation, and supporting human livelihoods. However, these ecosystems face numerous challenges that threaten their health, functionality, and capacity to provide essential services.

Addressing these challenges requires coordinated efforts, policy intervention, and community engagement.

Coastal Development and Habitat Loss

One of the primary threats to marine and coastal ecosystems is habitat loss due to urbanization, infrastructure projects, and land reclamation. Coastal development often involves clearing mangroves, dredging seagrass meadows, and altering tidal marshes to accommodate housing, tourism, and industry. These activities not only degrade ecosystems but also reduce their carbon sequestration potential and resilience to climate change impacts such as sea-level rise.

Habitat fragmentation caused by development disrupts ecological connectivity, limiting species movement and reducing genetic diversity. This fragmentation weakens ecosystems' ability to recover from disturbances, exacerbating their vulnerability.

Pollution and Contamination

Marine and coastal ecosystems are heavily impacted by pollution from various sources, including agricultural runoff, industrial waste, and plastic debris. Excessive nutrients from fertilizers cause eutrophication, leading to algal blooms that deplete oxygen levels in the water and create dead zones, where marine life cannot survive. Oil spills and heavy metals further contaminate coastal environments, affecting both aquatic organisms and human health.

Plastic pollution poses a pervasive threat to marine ecosystems. Microplastics ingested by marine species can accumulate through the food chain, disrupting biodiversity and potentially impacting human food security.

Overfishing and Unsustainable Practices

Overfishing and destructive fishing practices, such as trawling and the use of explosives, deplete fish populations and damage marine habitats. Seagrass meadows, coral reefs, and other critical habitats often bear the brunt of these activities, as they are physically disturbed or destroyed in the process. Unsustainable aquaculture practices also contribute to habitat degradation by increasing nutrient loads and introducing diseases to wild populations.

Climate Change Impacts

Climate change exacerbates existing threats to marine and coastal ecosystems. Rising sea levels inundate coastal habitats, while warming oceans affect species distributions, coral bleaching, and the productivity of seagrass meadows. Ocean acidification, caused by increased CO_2 absorption, disrupts the growth of calcifying organisms such as corals and shellfish, which are foundational to many marine ecosystems.

Extreme weather events, including hurricanes and typhoons, further stress these ecosystems by causing physical damage and altering salinity and sedimentation patterns.

Governance and Enforcement Gaps

Weak governance and inadequate enforcement of conservation regulations pose significant challenges to protecting marine and coastal ecosystems. In many regions, policies are insufficient or poorly implemented, allowing illegal activities such as unregulated fishing and habitat destruction to continue. Limited financial and technical resources further hinder the ability of authorities to monitor and protect these ecosystems effectively.

Community and Stakeholder Conflicts

Conflicts among stakeholders, including governments, industries, and local communities, often arise over competing interests in marine and coastal areas. Balancing conservation with economic development and resource use is complex, particularly in regions where communities rely on these ecosystems for their livelihoods.

Chapter 7: Policy Measures for Ecosystem-Based Mitigation

Ecosystem-based mitigation offers a powerful approach to addressing climate change by leveraging natural systems to capture and store carbon while providing additional environmental and socio-economic benefits. However, the success of such initiatives depends heavily on effective policy measures that create enabling environments for implementation.

This chapter examines the role of international frameworks, national climate strategies, and economic instruments in promoting ecosystem-based mitigation. It explores how carbon pricing, payments for ecosystem services, and programs like REDD+ incentivize conservation and restoration efforts. Additionally, the chapter highlights the importance of multi-stakeholder approaches and community engagement in ensuring equitable and sustainable outcomes. By linking policy measures to actionable solutions, this chapter underscores the critical role of governance and collaboration in advancing ecosystem-based mitigation at scale.

Overview of International Frameworks Supporting Ecosystem-Based Mitigation

Ecosystem-based mitigation, which harnesses natural systems to sequester carbon and address climate change, has gained recognition as a vital strategy for achieving global sustainability goals. A range of international frameworks underpins efforts to promote, implement, and scale up ecosystem-based mitigation, offering policy guidance, financial support, and collaborative opportunities. These frameworks provide a foundation for countries to align their strategies with global objectives and enhance the resilience of ecosystems and communities.

The Paris Agreement

The Paris Agreement, adopted in 2015 under the United Nations Framework Convention on Climate Change (UNFCCC), is a cornerstone of global climate governance. It emphasizes the importance of ecosystem-based mitigation by encouraging countries to include nature-based solutions in their NDCs. Many countries have incorporated forest conservation, reforestation, and wetland restoration into their NDCs as key strategies for achieving carbon neutrality.

The Paris Agreement also promotes financial mechanisms, such as the GCF, to support ecosystem-based initiatives in developing countries. By aligning national priorities with global goals, the agreement fosters collaboration and accountability in implementing ecosystem-based mitigation.

The Convention on Biological Diversity

The CBD recognizes the interdependence of biodiversity conservation and climate mitigation. Through its Aichi Biodiversity Targets (2010–2020) and the Kunming-Montreal Global Biodiversity Framework (2022), the CBD highlights the role of healthy ecosystems, such as forests, peatlands, and mangroves, in mitigating climate change.

Target 8 of the Kunming-Montreal Framework specifically addresses ecosystem-based approaches to climate action, aiming to enhance carbon sequestration while protecting biodiversity. The CBD's emphasis on ecosystem resilience underscores the need for integrated strategies that balance conservation, sustainable use, and climate mitigation.

The Ramsar Convention on Wetlands

The Ramsar Convention, established in 1971, is a pivotal framework for the conservation and sustainable use of wetlands. Wetlands, including peatlands, mangroves, and salt marshes, are critical blue carbon ecosystems with significant carbon sequestration potential.

The convention encourages member countries to designate and manage Ramsar Sites, wetlands of international importance, to preserve their ecological integrity and carbon storage capacity.

By promoting the wise use of wetlands, the Ramsar Convention aligns with broader climate and biodiversity objectives, enhancing the role of these ecosystems in global mitigation efforts.

REDD+ and Forest Conservation Initiatives

The Reducing Emissions from Deforestation and Forest Degradation (REDD+) mechanism, established under the UNFCCC, incentivizes countries to conserve and restore forests as carbon sinks. By providing financial rewards for reducing emissions from deforestation and forest degradation, REDD+ supports ecosystem-based mitigation while addressing socio-economic challenges in forest-dependent communities.

REDD+ also emphasizes co-benefits, such as biodiversity conservation and improved livelihoods, making it a comprehensive approach to sustainable development. International support for REDD+ includes funding from multilateral organizations, such as the World Bank's Forest Carbon Partnership Facility (FCPF) and the United Nations Collaborative Programme on Reducing Emissions from Deforestation and Forest Degradation in Developing Countries.

The SDGs

Adopted by the United Nations in 2015, the SDGs provide a broad framework for advancing ecosystem-based mitigation. Goals such as SDG 13 (Climate Action), SDG 15 (Life on Land), and SDG 14 (Life Below Water) emphasize the importance of protecting and restoring ecosystems to achieve sustainable development.

The SDGs encourage countries to integrate ecosystem-based approaches into national policies, fostering synergies between climate action, biodiversity conservation, and socio-economic well-being.

Integration of Ecosystem-Based Approaches into National Climate Policies

Ecosystem-based approaches are increasingly recognized as vital components of national climate policies due to their dual benefits of mitigating climate change and enhancing resilience to its impacts. These strategies leverage natural systems, such as forests, wetlands, and grasslands, to sequester carbon, regulate water cycles, and protect biodiversity. Integrating ecosystem-based approaches into national climate policies requires comprehensive planning, multi-sectoral collaboration, and alignment with international frameworks.

NDCs

Under the Paris Agreement, countries submit NDCs, outlining their commitments to reduce GHG emissions and adapt to climate change. Many nations have included ecosystem-based approaches in their NDCs, highlighting actions such as reforestation, wetland restoration, and sustainable land management. These initiatives reflect the growing recognition of ecosystems as cost-effective and scalable solutions for achieving climate goals.

For example, several tropical countries have pledged to reduce deforestation and restore degraded lands as part of their NDCs. By emphasizing ecosystem-based approaches, nations can demonstrate their commitment to global climate targets while addressing local environmental and socio-economic challenges.

Forest Conservation and Reforestation

Forests are central to ecosystem-based climate strategies, as they serve as significant carbon sinks. Integrating forest conservation and reforestation into national climate policies involves protecting existing forests, promoting afforestation, and restoring degraded forestlands. Many countries have adopted national programs aligned with global initiatives like REDD+ (Reducing Emissions from Deforestation and Forest Degradation) to incentivize forest conservation.

For instance, countries such as Brazil and Indonesia have implemented policies to curb illegal logging, establish protected areas, and promote sustainable forestry practices. These efforts not only reduce carbon emissions but also contribute to biodiversity conservation and support local livelihoods.

Wetland Protection and Restoration

Wetlands, including peatlands, mangroves, and tidal marshes, are among the most effective ecosystems for carbon sequestration. Their inclusion in national climate policies often focuses on rewetting drained wetlands, preventing land-use changes, and restoring degraded areas. Countries with significant wetland resources, such as the Democratic Republic of Congo and Malaysia, have integrated wetland conservation into their climate strategies to enhance carbon storage and reduce emissions.

Additionally, wetland protection contributes to climate adaptation by mitigating flooding, improving water quality, and protecting coastal communities from storm surges and sea-level rise. These co-benefits make wetlands a cornerstone of ecosystem-based approaches.

Sustainable Agriculture and Land Management

Agricultural policies that incorporate ecosystem-based approaches focus on practices like agroforestry, crop diversification, and soil conservation to reduce emissions and enhance carbon sequestration. Integrating these practices into national climate policies promotes sustainable land use while addressing food security challenges.

Countries such as Ethiopia and India have implemented large-scale programs to promote sustainable agriculture and restore degraded lands. These initiatives align with broader climate goals by reducing emissions from agriculture, increasing soil organic carbon, and improving ecosystem resilience.

Coastal and Marine Ecosystem Protection

Coastal and marine ecosystems, known as blue carbon ecosystems, are vital components of national climate policies for countries with extensive coastlines. Strategies include establishing MPAs, restoring mangroves, and protecting seagrass meadows. For example, Fiji and the Philippines have incorporated blue carbon initiatives into their climate action plans to enhance carbon sequestration and support fisheries.

Integrating blue carbon ecosystems into climate policies also strengthens coastal resilience, reducing vulnerability to climate-related impacts such as sea-level rise and extreme weather events.

Policy Instruments and Incentives

To facilitate the integration of ecosystem-based approaches, governments often employ policy instruments and economic incentives. Carbon pricing mechanisms, such as carbon taxes and emissions trading systems, encourage the adoption of ecosystem-based solutions by providing financial benefits for conservation and restoration efforts.

PES programs reward landowners and communities for maintaining ecosystems that deliver carbon sequestration and other services. For instance, Costa Rica's PES program incentivizes forest conservation and reforestation, contributing to its success in restoring forest cover and reducing emissions.

Challenges in Integration

Integrating ecosystem-based approaches into national climate policies is not without challenges. Limited financial resources, technical expertise, and institutional capacity can hinder implementation, particularly in developing countries. Additionally, competing land-use priorities, such as agriculture and urban development, often conflict with conservation goals.

To address these challenges, governments must prioritize capacity building, enhance inter-agency coordination, and secure funding from international sources such as the GCF. Multi-stakeholder engagement, including collaboration with local communities and the private sector, is also essential for ensuring equitable and sustainable outcomes.

Economic Instruments: Carbon Pricing, Payments for Ecosystem Services, and REDD

Economic instruments are essential tools for promoting ecosystem-based mitigation by incentivizing conservation and restoration activities while addressing the financial challenges of climate action. Carbon pricing, PES, and REDD+ are among the most effective mechanisms to align economic goals with environmental sustainability. These instruments encourage sustainable practices, mobilize financial resources, and create market-driven solutions to mitigate climate change.

Carbon Pricing: Incentivizing Climate Action

Carbon pricing is a market-based mechanism that assigns a monetary value to GHG emissions, creating an economic incentive to reduce them. The two primary forms of carbon pricing are carbon taxes and emissions trading systems (ETS), both of which encourage ecosystem-based approaches by making conservation and restoration economically viable.

Carbon Taxes

A carbon tax imposes a fixed fee on each ton of CO_2 or equivalent GHG emitted. By increasing the cost of emissions, this approach encourages businesses and individuals to adopt cleaner technologies and practices. Carbon taxes can also support ecosystem-based mitigation by directing revenue toward conservation projects, such as reforestation or wetland restoration.

For example, Colombia's carbon tax allocates a portion of its revenue to forest conservation initiatives, demonstrating how tax systems can simultaneously reduce emissions and enhance carbon sinks.

ETS

An ETS, also known as a cap-and-trade system, sets a limit on total emissions and allows entities to trade emission allowances within that cap. Entities that reduce emissions below their allocated limit can sell surplus allowances, creating a financial incentive for mitigation efforts.

Ecosystem-based projects, such as afforestation and soil carbon sequestration, can generate carbon credits that are tradable within an ETS. This creates a direct link between market dynamics and ecosystem conservation.

PES: Rewarding Conservation

PES are financial incentives provided to landowners, communities, or organizations for maintaining or enhancing ecosystems that deliver valuable services, such as carbon sequestration, water purification, and biodiversity conservation. PES schemes align economic benefits with environmental stewardship, making sustainable practices more appealing to stakeholders.

Key Features of PES Programs

• **Voluntary Participation**: Participants are incentivized but not required to adopt sustainable practices.

• **Conditional Payments**: Payments are contingent upon measurable conservation outcomes, such as maintaining forest cover or reducing soil erosion.

• **Targeted Services**: PES schemes focus on specific ecosystem services, such as carbon storage or water management.

Examples of PES in Action

Costa Rica's pioneering PES program has successfully incentivized forest conservation and reforestation. Landowners receive payments for maintaining forest cover, which supports carbon sequestration and biodiversity. The program is funded through a combination of government budgets and carbon offset markets, demonstrating the scalability of PES models.

REDD+: Combating Deforestation and Forest Degradation

REDD+ is an international mechanism under the United Nations Framework Convention on Climate Change (UNFCCC) that aims to reduce emissions from deforestation and forest degradation while promoting sustainable forest management and conservation. By providing financial incentives to developing countries, REDD+ encourages the preservation of forests as carbon sinks.

Core Components of REDD+

1. **Reducing Emissions**: Preventing deforestation and degradation to avoid the release of stored carbon.

2. **Enhancing Carbon Stocks**: Supporting afforestation, reforestation, and natural regeneration to increase forest carbon storage.

3. **Sustainable Management**: Promoting practices that balance forest conservation with economic needs.

Implementation and Benefits

Countries participating in REDD+ receive results-based payments for verified reductions in emissions. For instance, Brazil and Indonesia have received substantial funding through REDD+ for their efforts to curb deforestation and restore degraded lands. These payments help offset the opportunity costs of conservation, such as foregone revenue from agriculture or logging.

In addition to mitigating climate change, REDD+ delivers co-benefits, including biodiversity conservation, improved livelihoods for forest-dependent communities, and enhanced ecosystem resilience.

Challenges and Opportunities

While economic instruments hold significant potential, they also face challenges that require careful management and adaptation.

Challenges

• **Measurement and Verification**: Ensuring accurate measurement of carbon sequestration and other ecosystem services is complex and resource-intensive.

• **Equity and Inclusivity**: Economic instruments must balance the interests of various stakeholders, particularly marginalized communities, to avoid unintended social consequences.

• **Market Volatility**: Carbon pricing and credit markets can be influenced by economic fluctuations, affecting the stability of funding for ecosystem-based projects.

Opportunities

• **Scaling Up Funding**: Leveraging public-private partnerships and international cooperation can increase financial resources for ecosystem-based mitigation.

• **Technology Integration**: Advances in remote sensing, GIS, and blockchain technology enhance transparency and efficiency in implementing economic instruments.

• **Alignment with Global Goals**: Linking economic instruments to international frameworks, such as the Paris Agreement and the SDGs, strengthens their impact and scalability.

Multi-Stakeholder Approaches and Community Engagement

Ecosystem-based mitigation relies on collaborative efforts across various sectors and stakeholders to achieve meaningful and sustainable outcomes. Multi-stakeholder approaches and active community engagement are essential for addressing the complexity of climate change, ensuring equitable participation, and promoting the long-term success of conservation and restoration initiatives.

Importance of Multi-Stakeholder Collaboration

Multi-stakeholder approaches bring together governments, private sector actors, NGOs, academia, and local communities to design and implement ecosystem-based mitigation strategies. This collaborative framework ensures diverse perspectives, expertise, and resources are mobilized to address complex challenges. By aligning the interests of different stakeholders, these approaches create synergies that enhance the efficiency and scalability of mitigation efforts.

Governments play a key role in establishing supportive policies and regulatory frameworks, while the private sector contributes funding, innovation, and technical expertise. NGOs often act as intermediaries, facilitating dialogue among stakeholders and ensuring that marginalized groups are represented. Academic institutions provide research and data to guide decision-making, while local communities contribute invaluable knowledge about ecosystems and traditional practices.

The Role of Community Engagement

Community engagement is central to the success of ecosystem-based mitigation. Local communities are often the primary stewards of natural resources and possess a deep understanding of the ecosystems they depend on. Engaging these communities ensures that their knowledge and priorities are integrated into project planning and implementation.

Community participation fosters a sense of ownership and responsibility, increasing the likelihood of long-term sustainability. For example, involving communities in reforestation projects can enhance the success of tree planting initiatives by ensuring that species selection and land-use decisions align with local needs.

Equitable benefit-sharing is a critical component of community engagement. Providing economic incentives, such as PES or employment opportunities in conservation projects, ensures that communities derive tangible benefits from their involvement. This not only supports livelihoods but also builds trust and cooperation among stakeholders.

Case Examples of Multi-Stakeholder Approaches

A successful example of multi-stakeholder collaboration is the implementation of the REDD+ framework. REDD+ initiatives involve governments, international organizations, private companies, and local communities working together to protect forests while promoting sustainable development. Stakeholder consultations ensure that REDD+ strategies address local concerns and provide co-benefits such as biodiversity conservation and improved livelihoods.

In coastal regions, mangrove restoration projects often combine the efforts of NGOs, governments, and local communities. By integrating traditional knowledge with scientific techniques, these initiatives enhance the resilience of ecosystems and reduce vulnerability to climate impacts.

Challenges and Opportunities

Despite its benefits, multi-stakeholder collaboration and community engagement face challenges such as conflicting interests, power imbalances, and limited resources. Overcoming these barriers requires transparent communication, capacity-building efforts, and mechanisms for conflict resolution.

Digital tools, such as GIS and participatory mapping, can support collaboration by providing stakeholders with a shared understanding of ecosystem dynamics. Similarly, targeted funding and policy frameworks can incentivize inclusive participation and equitable decision-making.

Chapter 8: Monitoring and Measuring Ecosystem-Based Mitigation

Accurate monitoring and measurement are essential for evaluating the effectiveness of ecosystem-based mitigation strategies. Robust monitoring systems ensure that carbon sequestration goals are being met, provide insights into ecosystem health, and inform adaptive management practices. This chapter explores the tools and technologies used to monitor carbon sequestration, the development of indicators for measuring effectiveness, and the role of advanced techniques such as remote sensing and GIS.

The chapter also addresses the challenges of data availability, accuracy, and standardization, while highlighting opportunities for innovation and improvement. By emphasizing the importance of reliable data and measurement, this chapter underscores the critical role of monitoring in achieving sustainable and scalable ecosystem-based mitigation efforts.

Tools and Technologies for Monitoring Carbon Sequestration

Monitoring carbon sequestration is critical for assessing the effectiveness of ecosystem-based mitigation strategies. Advanced tools and technologies provide accurate measurements of carbon stored in terrestrial and aquatic ecosystems, enabling data-driven decision-making and accountability. These tools are increasingly essential as climate mitigation projects scale up, requiring robust systems to track progress and verify outcomes.

Remote Sensing Technologies

Remote sensing is one of the most widely used technologies for monitoring carbon sequestration. Satellite and aerial imagery provide large-scale, real-time data on land cover, vegetation health, and biomass. Sensors equipped on satellites, drones, or aircraft capture data across various wavelengths, enabling detailed analysis of ecosystems.

Key Remote Sensing Tools

• **LiDAR (Light Detection and Ranging):** LiDAR uses laser pulses to measure the height and density of vegetation. It is particularly effective in estimating above-ground biomass and carbon storage in forests.

• **Optical Sensors:** Satellites like Landsat and Sentinel-2 capture high-resolution optical imagery, which can be used to monitor vegetation cover, forest health, and land-use changes.

• **Radar Systems:** Synthetic Aperture Radar (SAR) can penetrate cloud cover and detect changes in vegetation structure, making it useful for monitoring tropical regions with frequent cloud cover.

Remote sensing provides comprehensive and repeatable data that supports the monitoring of carbon sequestration over time, offering insights into both successes and areas needing intervention.

Ground-Based Measurement Tools

Ground-based tools complement remote sensing by providing detailed, site-specific data on carbon storage in ecosystems. These methods are particularly useful for validating satellite-derived estimates and monitoring below-ground carbon.

Tree Measurements

In forests, carbon sequestration is often estimated by measuring tree attributes such as diameter, height, and species. These measurements are used to calculate biomass using established allometric equations, which estimate the carbon stored in individual trees and entire stands.

Soil Carbon Analysis

Soil carbon is a significant component of ecosystem-based mitigation, especially in grasslands, wetlands, and agricultural systems. Tools like soil corers and sensors collect soil samples or measure soil carbon content in situ. Laboratory analysis of soil samples provides precise information on organic and inorganic carbon levels, supporting strategies to enhance soil carbon storage.

Chambers for Gas Flux Measurement

To monitor carbon dynamics in wetlands, agricultural fields, and grasslands, gas flux chambers measure the exchange of CO_2 and CH_4 between the soil and the atmosphere. These measurements are crucial for understanding how management practices affect greenhouse gas emissions and carbon sequestration.

GIS

GIS integrates spatial and temporal data from various sources, enabling comprehensive analysis of carbon sequestration patterns. By overlaying remote sensing imagery, ground-based measurements, and environmental datasets, GIS provides valuable insights into ecosystem changes, land-use impacts, and carbon storage trends.

GIS tools allow users to map carbon sequestration hotspots, monitor deforestation, and plan restoration efforts. They also facilitate scenario modeling, enabling policymakers and stakeholders to evaluate the potential outcomes of different land-use strategies.

Carbon Accounting Software

Carbon accounting tools and software platforms are designed to quantify and report carbon sequestration data. These tools use algorithms to calculate carbon stocks and fluxes, integrating data from remote sensing, ground measurements, and GIS.

Examples of Carbon Accounting Tools

• **CBM-CFS3 (Carbon Budget Model of the Canadian Forest Sector):** Used for forest carbon accounting, this tool tracks carbon pools and fluxes over time.

• **COMET-Farm:** Developed for agricultural systems, this tool estimates carbon sequestration and greenhouse gas emissions based on management practices.

• **REDD+ Measurement, Reporting, and Verification (MRV) Systems:** These frameworks support developing countries in tracking carbon outcomes under REDD+ initiatives.

Emerging Technologies

Advancements in technology continue to enhance the precision and efficiency of monitoring carbon sequestration. Emerging tools include:

• **Artificial Intelligence (AI):** AI algorithms analyze vast datasets, identifying patterns and trends in carbon storage and land-use changes.

• **Blockchain Technology:** Blockchain ensures transparency and traceability in carbon markets, providing secure and verifiable records of carbon credits and sequestration outcomes.

• **IoT Sensors (Internet of Things):** Deployed in ecosystems, IoT sensors continuously monitor variables such as soil moisture, temperature, and gas flux, providing real-time data on carbon dynamics.

Challenges and Opportunities

Despite their effectiveness, tools and technologies for monitoring carbon sequestration face challenges, including high costs, data accessibility, and the need for technical expertise. Standardization of methods and protocols is also crucial for ensuring consistency and comparability across projects and regions.

Expanding the adoption of these tools requires investments in capacity building, funding for infrastructure, and fostering international collaboration. Technological advancements and integration of innovative solutions can further improve the accuracy and scalability of carbon monitoring systems.

Indicators for Measuring the Effectiveness of Ecosystem-Based Mitigation

Effective monitoring of ecosystem-based mitigation requires the use of robust indicators to measure progress and assess outcomes. These indicators serve as benchmarks for evaluating the success of conservation, restoration, and sustainable management efforts in achieving carbon sequestration and other environmental objectives. They also provide critical insights for adaptive management, ensuring that interventions remain aligned with climate goals and ecosystem health.

Carbon Sequestration Indicators

Carbon sequestration is a primary objective of ecosystem-based mitigation. Indicators for measuring carbon sequestration focus on quantifying carbon stocks and fluxes in various components of ecosystems, such as vegetation, soils, and sediments.

• **Above-Ground Biomass**: Measuring the carbon stored in trees, shrubs, and other vegetation provides an indicator of the carbon sequestration capacity of forests and agroforestry systems.

• **Below-Ground Biomass**: Root systems and soil organic matter are key reservoirs of carbon in grasslands, wetlands, and agricultural systems. Soil carbon content is often used as an indicator of below-ground carbon storage.

• **Net Primary Productivity (NPP)**: NPP measures the rate at which plants capture carbon through photosynthesis, reflecting the growth and carbon absorption potential of an ecosystem.

Ecosystem Health Indicators

Healthy ecosystems are more effective at sequestering carbon and providing co-benefits such as biodiversity conservation and water regulation. Indicators of ecosystem health help evaluate the resilience and functionality of ecosystems targeted by mitigation efforts.

• **Vegetation Cover**: The extent and density of vegetation cover indicate the health of forests, grasslands, and wetlands, as well as their capacity for carbon storage.

• **Biodiversity Indices**: Species richness and abundance are indicators of ecosystem stability and resilience, which are crucial for sustaining carbon sequestration over time.

• **Water Quality and Hydrology**: In wetlands and mangroves, water levels, salinity, and nutrient concentrations serve as indicators of ecosystem health and carbon storage potential.

Social and Economic Indicators

Ecosystem-based mitigation often aims to deliver socio-economic benefits alongside environmental outcomes. Measuring these benefits ensures that interventions support communities and align with sustainable development goals.

• **Livelihood Improvements**: Indicators such as income generation, employment rates, and access to resources measure the socio-economic benefits of projects like agroforestry or wetland restoration.

• **Community Participation**: The extent of community engagement and involvement in decision-making processes is a key indicator of the inclusiveness and sustainability of mitigation efforts.

• **Equity Metrics**: Indicators that track benefit-sharing and the inclusion of marginalized groups ensure that ecosystem-based mitigation efforts are socially equitable.

Climate Mitigation Performance Indicators

Climate-specific indicators evaluate the effectiveness of ecosystem-based mitigation in reducing GHG emissions and contributing to climate goals.

• **Greenhouse Gas Emission Reductions**: Measuring reductions in CO_2, CH_4, and nitrous oxide (N_2O) emissions provides a direct assessment of climate benefits.

• **Carbon Credit Generation**: The number of carbon credits generated through projects, such as those under REDD+ or carbon markets, indicates the scale of mitigation achieved.

Monitoring and Reporting Systems

Indicators are typically embedded within monitoring and reporting systems, such as MRV frameworks. These systems standardize data collection and reporting processes, ensuring consistency and transparency. Emerging technologies, including remote sensing, GIS, and IoT sensors, enhance the accuracy and scalability of indicator measurement.

Role of Remote Sensing and GIS in Tracking Ecosystem Changes

Remote sensing and GIS are indispensable tools for tracking ecosystem changes and supporting ecosystem-based mitigation strategies. These technologies provide comprehensive, real-time data on land cover, vegetation health, and carbon dynamics, enabling accurate monitoring of ecosystem conditions and informing decision-making processes. Their ability to cover large areas and integrate diverse datasets makes them especially valuable for assessing the impacts of conservation and restoration efforts.

Remote Sensing: Observing Ecosystems from Space and Air

Remote sensing involves the use of satellite, aerial, and drone-based sensors to collect data about the Earth's surface. This technology captures information across various wavelengths, including visible, infrared, and microwave spectra, allowing for detailed analysis of ecosystem attributes.

Applications in Ecosystem Monitoring

• **Land Cover and Vegetation Mapping**: Remote sensing enables the classification of ecosystems, such as forests, wetlands, and grasslands, and monitors changes over time. These insights help identify deforestation, habitat loss, and restoration progress.

• **Biomass and Carbon Estimation**: Sensors like LiDAR (Light Detection and Ranging) measure vegetation height and density, providing accurate estimates of above-ground biomass and carbon storage. Optical and radar sensors, such as those on the Landsat, Sentinel, and MODIS satellites, also contribute to tracking vegetation growth and carbon fluxes.

• **Water Quality and Hydrology**: In coastal and wetland ecosystems, remote sensing assesses water quality parameters like turbidity, chlorophyll content, and salinity. These metrics are critical for monitoring the health of mangroves, seagrass meadows, and tidal marshes.

Advantages of Remote Sensing

Remote sensing offers scalability, allowing for the monitoring of vast and inaccessible regions. Its repeatable observations provide consistent datasets for detecting trends and assessing the effectiveness of ecosystem-based mitigation efforts. Additionally, the integration of machine learning algorithms enhances the ability to analyze large datasets and identify patterns.

GIS: Integrating Spatial Data

GIS is a powerful tool for managing, analyzing, and visualizing spatial data. It integrates data from remote sensing, ground-based measurements, and other sources to create detailed maps and models of ecosystems.

Applications in Ecosystem-Based Mitigation

• **Mapping and Planning**: GIS creates detailed land-use and vegetation maps that guide the planning of conservation and restoration projects. These maps help identify areas with high carbon sequestration potential or vulnerable ecosystems in need of protection.

• **Monitoring and Evaluation**: By overlaying temporal datasets, GIS tracks changes in land cover, biodiversity, and carbon storage. This capability supports the evaluation of mitigation project outcomes and ensures accountability in reporting.

• **Scenario Modeling**: GIS allows stakeholders to simulate the effects of various land-use scenarios, such as reforestation or wetland restoration, helping prioritize actions that maximize carbon sequestration and ecosystem resilience.

Advantages of GIS

GIS enhances collaboration by integrating data from multiple stakeholders and disciplines into a shared platform. Its visualization capabilities, such as heat maps and 3D modeling, make complex datasets accessible and actionable for policymakers, researchers, and practitioners.

Challenges and Opportunities

While remote sensing and GIS are powerful tools, they face challenges such as data accessibility, high costs, and technical expertise requirements. Advancements in open-source platforms, such as Google Earth Engine and QGIS, are addressing these barriers by providing free or affordable access to tools and datasets. Innovations in sensor technology and cloud computing are further enhancing the precision and scalability of these technologies.

Data Challenges and Opportunities for Improvement

Data is essential for the successful implementation and monitoring of ecosystem-based mitigation strategies. However, challenges related to data collection, accuracy, accessibility, and integration often hinder effective decision-making. Addressing these challenges while leveraging emerging technologies and collaborative opportunities can significantly enhance data quality and usability.

Challenges in Data Collection and Accuracy

• **Data Gaps**: In many regions, particularly in developing countries, data on ecosystems and carbon sequestration is incomplete or unavailable. These gaps limit the ability to assess ecosystem health and track progress in mitigation efforts.

• **Measurement Uncertainty**: Variability in methodologies and tools for data collection can lead to inconsistent results. For instance, discrepancies in remote sensing technologies and ground-based measurements may affect the accuracy of carbon estimates.

• **Temporal and Spatial Resolution**: Limited temporal or spatial resolution in datasets can obscure critical trends. High-resolution data is often unavailable or prohibitively expensive, especially for large-scale projects.

• **Data Fragmentation**: Ecosystem data is often siloed across institutions, sectors, and countries. This fragmentation hinders comprehensive analysis and the ability to implement integrated mitigation strategies.

Challenges in Data Accessibility and Integration

• **Restricted Access**: Proprietary datasets and high licensing costs limit access to valuable ecosystem data. Open data initiatives remain insufficient in providing global coverage.

• **Lack of Standardization**: Variations in data formats and collection protocols complicate integration and comparison across regions and projects.

• **Technical Expertise**: Many organizations lack the capacity or expertise to analyze and interpret complex datasets, limiting their ability to leverage available information effectively.

Opportunities for Improvement

• **Advancements in Technology**: Emerging technologies, such as drones, IoT sensors, and artificial intelligence, offer new ways to collect high-resolution, real-time data. These tools enhance the accuracy and scalability of monitoring efforts.

• **Open Data Platforms**: Initiatives like Google Earth Engine and Global Forest Watch provide free access to satellite imagery and analytical tools, improving data accessibility for stakeholders worldwide.

• **Standardization Efforts**: Developing global standards for data collection, processing, and reporting can reduce inconsistencies and improve the reliability of ecosystem data.

• **Capacity Building**: Training programs and partnerships can enhance technical expertise, enabling organizations to utilize advanced data tools effectively.

• **Collaborative Networks**: International collaborations and data-sharing agreements can bridge data gaps, promote transparency, and foster integrated approaches to ecosystem-based mitigation.

Chapter 9: Overcoming Barriers to Implementation

Despite the significant potential of ecosystem-based mitigation to address climate change and deliver co-benefits such as biodiversity conservation and improved livelihoods, its implementation faces numerous barriers. These challenges include financial constraints, institutional inefficiencies, technical limitations, and social factors that hinder the adoption and scaling of solutions.

This chapter explores the common barriers to implementing ecosystem-based mitigation strategies and examines approaches to address them effectively. It highlights the importance of capacity building, knowledge sharing, and fostering public-private partnerships to overcome financial and technical challenges. Additionally, the chapter emphasizes the role of community engagement in addressing social and cultural barriers, ensuring that solutions are equitable and sustainable. By drawing on examples of successful implementation, this chapter provides actionable insights for overcoming obstacles and advancing ecosystem-based mitigation efforts at scale.

Common Barriers: Financial, Institutional, Technical, and Social

Implementing ecosystem-based mitigation faces a variety of barriers that limit its effectiveness and scalability. These challenges arise from financial, institutional, technical, and social constraints, which often intersect and require coordinated efforts to address. Overcoming these barriers is essential for realizing the full potential of ecosystem-based approaches in mitigating climate change.

Financial Barriers

One of the most significant barriers to implementing ecosystem-based mitigation is the lack of adequate financial resources.

• **High Initial Costs**: Many ecosystem restoration and conservation projects require significant upfront investment, such as reforestation initiatives, wetland restoration, or mangrove replanting. These costs can deter governments, private investors, and local communities from initiating such projects.

• **Limited Long-Term Funding**: Sustaining ecosystem-based projects requires ongoing funding for maintenance, monitoring, and adaptive management. However, long-term financial commitments are often lacking, leaving projects vulnerable to failure.

• **Market Failures**: Ecosystem services, such as carbon sequestration, water filtration, and biodiversity conservation, are often undervalued in economic systems. This undervaluation leads to insufficient incentives for stakeholders to invest in ecosystem-based approaches.

Addressing financial barriers requires innovative funding mechanisms, such as carbon credit markets, PES, and blended finance models that leverage public and private resources.

Institutional Barriers

Weak institutional frameworks and governance issues pose significant challenges to ecosystem-based mitigation.

• **Policy Gaps**: In many countries, policies supporting ecosystem-based approaches are either absent or insufficiently detailed. This lack of clear guidance hinders the integration of these strategies into broader climate and development plans.

• **Coordination Challenges**: Ecosystem-based mitigation often involves multiple sectors, such as forestry, agriculture, and water management. Poor coordination among government agencies and stakeholders can lead to fragmented efforts and reduced effectiveness.

• **Enforcement Weaknesses**: Even where policies exist, enforcement is often inadequate due to limited institutional capacity, corruption, or competing land-use priorities. This undermines conservation efforts and creates opportunities for unsustainable practices.

Strengthening governance structures, fostering inter-agency collaboration, and building institutional capacity are critical to overcoming these barriers.

Technical Barriers

Technical challenges, including gaps in knowledge, tools, and expertise, can impede the implementation of ecosystem-based mitigation.

• **Knowledge Gaps**: Many stakeholders lack an understanding of how ecosystem-based approaches work or their potential co-benefits. Limited awareness among policymakers, land managers, and local communities reduces adoption rates.

• **Data and Monitoring Limitations**: Reliable data on ecosystem health, carbon sequestration, and land-use changes are often unavailable or difficult to access. This hinders the ability to plan, implement, and evaluate projects effectively.

• **Capacity Constraints**: The technical expertise needed to design and implement ecosystem-based mitigation strategies is often lacking, particularly in developing countries. Challenges in applying advanced technologies, such as remote sensing and GIS, further limit effectiveness.

Investing in research, capacity building, and technological innovation is necessary to address these technical barriers and improve implementation outcomes.

Social Barriers

Social and cultural factors can also hinder the adoption of ecosystem-based mitigation, particularly at the community level.

• **Resistance to Change**: Communities and stakeholders may resist ecosystem-based projects due to a lack of understanding, mistrust of external actors, or concerns about losing access to natural resources.

• **Equity and Inclusion**: Projects that do not adequately involve marginalized groups, such as indigenous peoples and women, risk perpetuating inequalities and reducing the social acceptability of mitigation efforts.

• **Conflicts Over Land Use**: Competing demands for land, such as agriculture, urbanization, and infrastructure development, can create conflicts that delay or derail ecosystem-based initiatives.

Addressing social barriers requires active community engagement, equitable benefit-sharing mechanisms, and participatory decision-making processes that build trust and inclusivity.

Capacity Building and Knowledge-Sharing Initiatives

Capacity building and knowledge-sharing initiatives are critical components of successful ecosystem-based mitigation strategies. These efforts empower individuals, communities, and institutions with the skills, knowledge, and resources needed to implement, manage, and sustain conservation and restoration projects effectively. By addressing gaps in technical expertise and fostering collaborative learning, these initiatives enhance the scalability and long-term success of ecosystem-based approaches.

The Role of Capacity Building

Capacity building focuses on developing the technical and institutional capabilities required for implementing ecosystem-based mitigation. This includes training individuals in specialized skills, strengthening organizations' operational capacities, and enhancing the overall governance of mitigation initiatives.

Key Areas of Focus:

• **Technical Training**: Providing training on advanced tools and techniques, such as remote sensing, GIS, and carbon accounting, equips stakeholders with the knowledge needed to monitor and measure ecosystem changes accurately.

• **Institutional Strengthening**: Building institutional capacity involves improving organizational structures, processes, and resources to support policy implementation, enforcement, and project management.

• **Policy Development**: Capacity-building programs often include support for policymakers in drafting and refining legislation that integrates ecosystem-based mitigation into broader climate and development plans.

Knowledge Sharing as a Catalyst for Success

Knowledge-sharing initiatives create opportunities for stakeholders to exchange information, experiences, and best practices. These exchanges foster innovation and build a collective understanding of effective ecosystem-based mitigation strategies.

Mechanisms for Knowledge Sharing:

• **Workshops and Conferences**: Organized events bring together diverse stakeholders, including researchers, practitioners, and policymakers, to share insights, discuss challenges, and explore collaborative solutions.

• **Online Platforms**: Digital platforms, such as webinars, knowledge hubs, and forums, provide accessible spaces for stakeholders to access and disseminate information on ecosystem-based mitigation.

• **Community-Based Learning**: Local knowledge-sharing networks enable communities to share traditional practices and context-specific solutions that contribute to ecosystem restoration and conservation.

Collaborative Initiatives

Collaborative capacity-building efforts often involve partnerships between governments, NGOs, academia, and the private sector. These partnerships leverage resources and expertise to deliver impactful programs.

• **Global Partnerships**: Initiatives like the GEF and the GCF provide funding and technical support for capacity-building programs in developing countries.

• **Regional Networks**: Regional initiatives, such as the African Forest Landscape Restoration Initiative (AFR100), promote knowledge sharing and capacity building across countries with similar ecological and socio-economic contexts.

Challenges and Opportunities

While capacity building and knowledge-sharing initiatives are essential, they face challenges such as limited funding, insufficient infrastructure, and language barriers. Addressing these challenges requires sustained investment, the use of innovative digital tools, and efforts to tailor programs to local needs.

Emerging opportunities include integrating AI and machine learning to enhance knowledge dissemination and improve training effectiveness. Additionally, expanding open-access platforms ensures that resources and information are widely available to stakeholders at all levels.

Role of Public-Private Partnerships in Scaling Up Solutions

Public-private partnerships (PPPs) play a crucial role in scaling up ecosystem-based mitigation solutions by combining the strengths of government agencies, private companies, and other stakeholders. These collaborations leverage financial resources, technical expertise, and innovation to address barriers and create scalable, sustainable solutions to combat climate change.

Mobilizing Financial Resources

One of the key contributions of PPPs is their ability to mobilize substantial financial resources. Governments often face budgetary constraints in funding large-scale ecosystem restoration and conservation projects. By partnering with private entities, they can access additional capital to support these initiatives.

Examples of Financial Collaboration:

• Private companies can invest in ecosystem restoration projects through corporate social responsibility (CSR) initiatives or offset programs.

• Governments can offer incentives, such as tax breaks or grants, to encourage private investment in ecosystem-based mitigation.

• Blended finance models, which combine public funding with private investment, reduce risks and attract capital for high-impact projects.

These mechanisms not only ensure adequate funding but also promote long-term sustainability by aligning private sector goals with public climate objectives.

Harnessing Technical Expertise and Innovation

Private sector involvement brings advanced technologies, research capabilities, and innovation to ecosystem-based mitigation projects.

Key Contributions:

• **Technology Development**: Companies specializing in remote sensing, GIS, and AI contribute cutting-edge tools for monitoring ecosystem changes and carbon sequestration.

• **Efficient Practices**: Private sector expertise in sustainable agriculture, forestry, and aquaculture can optimize resource use and minimize environmental impact.

• **Capacity Building**: Through training programs and knowledge-sharing initiatives, private organizations can enhance the technical capabilities of local communities and government agencies.

This integration of expertise and innovation accelerates the implementation and effectiveness of ecosystem-based solutions.

Enhancing Governance and Accountability

PPPs improve governance and accountability in ecosystem-based mitigation projects by fostering transparency and shared responsibility.

• **Shared Goals**: Public and private stakeholders collaborate to define clear objectives, ensuring alignment with climate targets and local priorities.

• **Monitoring and Reporting**: Private companies often implement robust monitoring systems, ensuring data accuracy and compliance with environmental standards.

• **Stakeholder Engagement**: PPPs facilitate collaboration with local communities and NGOs, ensuring that projects are inclusive and socially equitable.

By enhancing governance structures, PPPs contribute to the long-term success of mitigation initiatives.

Challenges and Opportunities

While PPPs offer significant benefits, they also face challenges such as misaligned priorities, limited trust, and regulatory barriers. Addressing these challenges requires clear agreements, transparent communication, and supportive policy frameworks.

Emerging opportunities include integrating sustainability into corporate strategies and leveraging global initiatives like the GCF and REDD+, which encourage private sector participation in ecosystem-based mitigation.

Case Examples of Successful Overcoming of Barriers

Overcoming barriers to ecosystem-based mitigation requires innovative approaches, collaboration, and persistent efforts. Successful case examples demonstrate how financial, institutional, technical, and social challenges can be addressed to achieve meaningful climate outcomes while delivering co-benefits.

Restoration of Degraded Forests in Costa Rica

Costa Rica provides a leading example of overcoming financial and institutional barriers to ecosystem restoration. Faced with extensive deforestation in the 20th century, the country introduced a PES program in the 1990s. Funded through fuel taxes and carbon credit sales, the program incentivized landowners to conserve and restore forests.

By creating a strong institutional framework and aligning financial incentives with conservation goals, Costa Rica not only reversed deforestation but also enhanced biodiversity, water resources, and carbon sequestration. The program's success highlights the importance of policy innovation and public-private collaboration in overcoming resource constraints.

Mangrove Restoration in the Philippines

In the Philippines, community-driven mangrove restoration projects have successfully addressed social and technical barriers. Mangroves, critical for coastal protection and carbon sequestration, faced degradation due to aquaculture expansion and urban development.

Collaborative efforts between local communities, NGOs, and government agencies implemented restoration projects that included technical training for mangrove planting and maintenance. Community members were actively engaged in decision-making and received economic benefits, such as employment opportunities and improved fisheries.

The inclusion of local knowledge and equitable benefit-sharing ensured the long-term sustainability of these initiatives, emphasizing the value of participatory approaches.

Wetland Protection through REDD+ in the Democratic Republic of Congo

The Democratic Republic of Congo (DRC) utilized the REDD+ framework to address deforestation and degradation in its vast peatlands. Financial and technical support from international organizations enabled the DRC to map its peatland resources and develop strategies for sustainable management.

Capacity-building initiatives provided local stakeholders with tools and knowledge for monitoring peatland health and implementing conservation practices. The project demonstrated how international cooperation and funding mechanisms can overcome technical and financial barriers while protecting globally significant carbon sinks.

Chapter 10: The Future of Ecosystem-Based Mitigation

Ecosystem-based mitigation holds immense potential to address climate change while delivering co-benefits such as biodiversity conservation, enhanced livelihoods, and ecosystem resilience. As global challenges intensify, innovative approaches and forward-thinking strategies will be essential to scale up these solutions effectively. This chapter explores emerging trends and technologies in ecosystem-based mitigation, including advancements in monitoring tools, financial mechanisms, and policy frameworks.

The chapter also examines the role of ecosystems in achieving global climate goals, such as the Paris Agreement targets, and discusses pathways to mainstream nature-based solutions into development planning. By envisioning a sustainable and resilient future, this chapter highlights the critical role of collaboration, innovation, and integrated approaches in shaping the next phase of ecosystem-based mitigation efforts.

Emerging Trends and Innovations in Ecosystem-Based Mitigation

Ecosystem-based mitigation is evolving as an essential strategy to combat climate change while addressing biodiversity loss and fostering sustainable development. Advancements in technology, finance, and policy are driving new trends and innovations, making these approaches more effective, scalable, and inclusive. This section explores emerging developments that are shaping the future of ecosystem-based mitigation.

Technological Advancements in Monitoring and Assessment

The integration of cutting-edge technologies is transforming the way ecosystem-based mitigation efforts are monitored and evaluated.

• **Remote Sensing and AI:** Satellite imagery, drones, and AI-powered analytics enable precise monitoring of ecosystem changes and carbon dynamics. Technologies such as LiDAR (Light Detection and Ranging) assess forest biomass, while machine learning models analyze data trends to improve decision-making. These tools enhance the accuracy and efficiency of monitoring efforts, particularly in remote or inaccessible areas.

• **GIS:** GIS platforms integrate spatial data to model ecosystem functions, identify restoration priorities, and assess mitigation outcomes. Emerging applications include 3D ecosystem mapping and predictive modeling of climate impacts on ecosystems.

• **Internet of Things (IoT) Sensors:** IoT devices measure environmental variables like soil moisture, carbon flux, and water levels in real time. These sensors provide actionable insights for adaptive management of forests, wetlands, and agricultural landscapes.

Innovative Financial Mechanisms

New financing models are being developed to address the funding challenges associated with ecosystem-based mitigation.

• **Carbon Markets and Credits:** The expansion of voluntary and compliance carbon markets provides opportunities to fund mitigation projects. Blue carbon ecosystems, such as mangroves and seagrass meadows, are increasingly being recognized in these markets, creating financial incentives for their conservation and restoration.

• **Green Bonds and Blended Finance:** Green bonds and blended finance models mobilize public and private capital for large-scale ecosystem restoration projects. These mechanisms reduce investment risks and attract diverse funding sources.

• **PES:** Innovations in PES programs, such as blockchain-based transactions, enhance transparency and trust in financial flows. These systems reward landowners and communities for maintaining ecosystems that provide climate and environmental benefits.

Policy and Governance Innovations

Policy frameworks are becoming more integrated and inclusive, supporting the mainstreaming of ecosystem-based mitigation into national and international strategies.

• **Nature-Positive Policies:** Governments are adopting nature-positive approaches that align economic development with conservation goals. Policies promoting agroforestry, wetland restoration, and sustainable fisheries are examples of such integration.

• **Incentives for Private Sector Engagement:** Governments are increasingly offering tax breaks, subsidies, and other incentives to encourage private sector participation in ecosystem-based mitigation. This fosters collaboration between public and private stakeholders, accelerating implementation.

• **Regional and Global Initiatives:** Frameworks like the Kunming-Montreal Global Biodiversity Framework and REDD+ are aligning climate, biodiversity, and sustainable development objectives. These initiatives promote shared accountability and resource mobilization across nations.

Community-Centered Approaches

Innovative approaches prioritize the inclusion and empowerment of local communities, recognizing their role as key stewards of ecosystems.

• **Participatory Decision-Making:** Community-led restoration projects, supported by capacity-building programs, foster local ownership and ensure alignment with socio-economic needs. Engaging communities in planning and implementation enhances sustainability.

• **Equity and Inclusion:** Emerging frameworks focus on ensuring equitable benefit-sharing, particularly for marginalized groups, including indigenous peoples and women. These approaches address historical inequities while maximizing community participation.

• **Livelihood Integration:** Combining ecosystem-based mitigation with income-generating activities, such as sustainable agriculture and ecotourism, strengthens community resilience and incentivizes conservation efforts.

Integration of Ecosystem Services into Development Planning

Mainstreaming ecosystem-based mitigation into broader development agendas ensures long-term impact and scalability.

• **Climate-Resilient Infrastructure:** Planners are increasingly integrating natural infrastructure, such as mangroves for coastal protection or wetlands for flood mitigation, into urban and rural development strategies. These solutions provide cost-effective alternatives to traditional infrastructure while offering ecological benefits.

• **Sustainable Agriculture and Food Systems:** Innovations in sustainable farming practices, such as regenerative agriculture and agroforestry, enhance carbon sequestration while addressing food security challenges. These practices also reduce agricultural emissions, contributing to climate mitigation.

Collaborative Knowledge Networks

Emerging knowledge-sharing platforms and networks are fostering global collaboration and innovation.

• **Open-Access Platforms:** Digital hubs like the Global Forest Watch and Google Earth Engine provide stakeholders with free access to data, tools, and best practices for ecosystem-based mitigation. These resources enhance transparency and democratize knowledge.

• **South-South Cooperation:** Developing countries are increasingly collaborating to share lessons learned and successful practices. This peer-to-peer knowledge exchange strengthens capacity and promotes context-specific solutions.

• **Public-Private Research Collaborations:** Partnerships between academic institutions, governments, and private companies drive innovation by combining research expertise with practical applications.

Role of Ecosystems in Achieving Global Climate Goals

Ecosystems play a pivotal role in achieving global climate goals by serving as natural carbon sinks, enhancing resilience to climate impacts, and providing critical ecosystem services. Forests, wetlands, grasslands, and coastal ecosystems contribute significantly to climate mitigation and adaptation efforts, making their protection and restoration essential for meeting international climate targets such as those outlined in the Paris Agreement and the SDGs.

Carbon Sequestration and Storage

Ecosystems are central to global efforts to mitigate climate change through their capacity to capture and store CO_2 from the atmosphere. Forests, for example, sequester significant amounts of carbon in their biomass and soils, accounting for about 30% of global CO_2 emissions absorbed annually. Similarly, wetlands, including peatlands and mangroves, store vast amounts of carbon in their soils, often exceeding the carbon storage capacity of terrestrial ecosystems.

Coastal ecosystems, such as seagrass meadows and tidal salt marshes, contribute to blue carbon sequestration by capturing carbon in sediments and vegetation. Protecting these ecosystems prevents the release of stored carbon while enhancing their ability to act as carbon sinks. Their role in reducing GHG emissions is integral to achieving net-zero targets.

Enhancing Climate Resilience

Beyond carbon sequestration, ecosystems play a critical role in helping communities and landscapes adapt to climate change. Wetlands and mangroves act as natural buffers against extreme weather events, reducing the impacts of storm surges, floods, and coastal erosion. Forests regulate local climates by moderating temperatures, maintaining water cycles, and reducing the risk of drought.

Healthy ecosystems also enhance biodiversity, which underpins resilience by supporting a variety of species that contribute to ecosystem functionality. Biodiverse ecosystems are better able to adapt to changing environmental conditions, ensuring their continued provision of essential services in the face of climate stressors.

Supporting SDGs

Ecosystems contribute directly to several SDGs, particularly those related to climate action (SDG 13), life on land (SDG 15), and life below water (SDG 14). Ecosystem-based approaches to climate mitigation and adaptation align with these goals by integrating environmental, social, and economic objectives.

Ecosystems also support livelihoods, particularly in rural and indigenous communities that depend on natural resources. Sustainable management of ecosystems provides opportunities for income generation, such as through ecotourism, sustainable agriculture, and carbon credit markets. These benefits reinforce the social and economic dimensions of global climate goals.

Global Frameworks and Ecosystem Integration

International frameworks emphasize the role of ecosystems in climate strategies. The Paris Agreement encourages countries to include nature-based solutions in their NDCs. Many countries have committed to reforestation, wetland restoration, and sustainable land-use practices as part of their climate action plans.

Initiatives like REDD+ provide financial incentives for protecting forests, aligning conservation goals with carbon reduction targets. Similarly, the Kunming-Montreal Global Biodiversity Framework highlights the interconnectedness of biodiversity conservation and climate action, promoting ecosystem-based solutions to address global challenges.

Challenges to Ecosystem Protection

While ecosystems are vital for achieving climate goals, they face significant threats from deforestation, land degradation, pollution, and climate change itself. Habitat loss and ecosystem degradation reduce carbon sequestration capacity and increase vulnerability to climate impacts. Addressing these challenges requires coordinated efforts to enforce protection measures, restore degraded areas, and integrate ecosystem-based approaches into broader development agendas.

Opportunities for Scaling Up Ecosystem-Based Solutions

Advancements in technology, such as remote sensing and GIS, are improving the ability to monitor ecosystem health and track carbon sequestration. Economic instruments, including PES and carbon credit markets, are creating incentives for conservation and restoration efforts.

Collaborative initiatives at the regional and global levels, supported by frameworks like the GCF and the GEF, are scaling up investments in ecosystem-based solutions. These efforts provide critical funding and technical expertise to maximize the role of ecosystems in achieving climate goals.

Pathways for Mainstreaming Nature-Based Solutions in Development Planning

Mainstreaming NbS into development planning is essential to address climate change, enhance ecosystem resilience, and achieve sustainable development goals. Integrating NbS into policies, programs, and infrastructure projects ensures that natural systems are leveraged to mitigate climate impacts, support biodiversity, and provide socio-economic benefits. This section explores key pathways for embedding NbS into development planning at local, national, and global levels.

Incorporating NbS into Policy Frameworks

Effective integration of NbS begins with embedding these approaches into policy and regulatory frameworks. Governments can create enabling environments that prioritize the use of natural systems in planning and decision-making.

• **National Strategies and Action Plans**: Incorporating NbS into climate action strategies, such as NDCs under the Paris Agreement, ensures alignment with global climate goals. Countries can commit to reforestation, wetland restoration, and agroforestry as part of their climate mitigation and adaptation plans.

• **Land-Use Planning**: Integrating NbS into zoning laws, land-use policies, and urban planning frameworks ensures that development projects minimize ecosystem disruption and enhance natural infrastructure. Policies can mandate the preservation of green corridors, buffer zones, and natural floodplains.

• **Sectoral Policies**: Mainstreaming NbS across sectors such as agriculture, forestry, and water management enhances their impact. For instance, promoting agroforestry in agricultural policies improves carbon sequestration and soil health.

Embedding NbS in Infrastructure Development

Natural infrastructure offers cost-effective and sustainable alternatives to traditional engineered solutions. Mainstreaming NbS into infrastructure planning can enhance resilience and provide multiple co-benefits.

• **Urban Green Infrastructure**: Cities can incorporate NbS into urban design through green roofs, permeable pavements, and urban forests. These features reduce heat island effects, manage stormwater, and improve air quality.

• **Coastal Protection**: Coastal development projects can integrate mangroves, seagrass meadows, and coral reefs to reduce wave energy, prevent erosion, and protect communities from storm surges.

• **Integrated Water Management**: NbS such as wetland restoration and riparian buffers can be incorporated into water resource management to improve water quality, recharge aquifers, and mitigate flooding.

Leveraging Economic Incentives

Economic tools and incentives play a critical role in encouraging the adoption of NbS in development planning.

• **PES**: PES programs reward stakeholders for maintaining ecosystems that deliver services such as carbon sequestration, water purification, and disaster risk reduction.

• **Carbon Markets**: Incorporating blue and green carbon projects into carbon trading schemes provides financial incentives for conserving and restoring ecosystems.

• **PPPs**: Collaborations between governments, businesses, and NGOs can mobilize funding and technical expertise for NbS projects. For example, companies investing in carbon offsets can fund reforestation or wetland restoration initiatives.

Enhancing Technical Capacity and Knowledge Sharing

Building capacity and fostering knowledge-sharing networks are essential to mainstream NbS effectively.

• **Training Programs**: Providing training for planners, engineers, and policymakers on the design and implementation of NbS ensures that these approaches are integrated into development projects.

• **Research and Innovation**: Advancing scientific research on the effectiveness and scalability of NbS informs evidence-based decision-making. Innovations in remote sensing, GIS, and AI enhance the monitoring and evaluation of NbS.

• **Knowledge Hubs and Networks**: Platforms such as the Nature-Based Solutions Initiative and Global Forest Watch facilitate information exchange, showcasing best practices and successful case studies.

Engaging Stakeholders and Communities

Inclusive and participatory approaches are critical for the successful integration of NbS in development planning. Engaging diverse stakeholders ensures that NbS align with local needs and priorities.

• **Community-Led Initiatives**: Empowering local communities to lead NbS projects fosters ownership and long-term sustainability. Participatory planning processes allow communities to contribute traditional knowledge and co-develop solutions.

• **Private Sector Involvement**: Encouraging businesses to adopt NbS in their operations and supply chains accelerates their integration into economic systems. Companies in agriculture, energy, and construction can play a pivotal role in scaling NbS.

• **Cross-Sector Collaboration**: Partnerships among governments, NGOs, academic institutions, and the private sector ensure that NbS are implemented effectively and equitably.

Monitoring and Evaluation

Integrating robust monitoring and evaluation frameworks into NbS projects ensures accountability and guides adaptive management.

• **Indicators and Metrics**: Developing standardized indicators to measure the performance of NbS projects, such as carbon sequestration, biodiversity enhancement, and socio-economic benefits, supports transparent reporting.

• **Technology Integration**: Using remote sensing, GIS, and IoT sensors for data collection and analysis improves the accuracy and efficiency of monitoring systems.

• **Feedback Mechanisms**: Incorporating feedback loops into planning processes allows stakeholders to refine NbS interventions based on lessons learned.

Mainstreaming nature-based solutions into development planning is essential for achieving climate resilience, biodiversity conservation, and sustainable development. By embedding NbS into policies, infrastructure projects, and economic systems, and fostering collaboration among stakeholders, these solutions can be scaled effectively. Addressing capacity gaps, leveraging technology, and promoting inclusive decision-making will ensure that NbS become a cornerstone of global development strategies.

Vision for a Sustainable and Resilient Future

A sustainable and resilient future hinges on the integration of ecosystem-based approaches into global strategies for climate mitigation, adaptation, and sustainable development. This vision encompasses thriving ecosystems, empowered communities, and robust governance systems working together to address the challenges of climate change and environmental degradation. Achieving this future requires a commitment to leveraging nature's potential while ensuring equitable benefits for present and future generations.

Thriving Ecosystems as Climate Solutions

At the heart of a sustainable future lies the preservation and restoration of ecosystems that act as natural climate solutions. Forests, wetlands, grasslands, and coastal ecosystems must be protected to maintain their critical roles as carbon sinks and biodiversity hotspots. Large-scale reforestation, afforestation, and wetland restoration projects will enhance carbon sequestration, mitigate greenhouse gas emissions, and create habitats for diverse species.

Blue carbon ecosystems, such as mangroves, seagrass meadows, and coral reefs, will serve as vital buffers against climate impacts, protecting coastal communities from storm surges and sea-level rise. The integration of nature-based solutions into urban landscapes— through green roofs, urban forests, and permeable infrastructure—

will address urban heat islands and improve the quality of life for city dwellers.

Empowered Communities at the Core of Solutions

A resilient future is one where communities are central to decision-making and benefit equitably from ecosystem-based mitigation and adaptation efforts. Local populations, especially indigenous peoples and rural communities, possess invaluable knowledge of their environments. Their active participation in designing and implementing ecosystem-based projects ensures that solutions are context-specific and sustainable.

Livelihood opportunities generated by conservation and restoration projects, such as ecotourism, sustainable farming, and PES, will enhance economic resilience while fostering environmental stewardship. Equitable benefit-sharing mechanisms will address social disparities and ensure that marginalized groups, including women and indigenous peoples, are not left behind.

Innovation and Technology for Sustainable Progress

Technological advancements will play a key role in realizing a sustainable future. Remote sensing, GIS, and AI will enable precise monitoring of ecosystems, while IoT sensors will provide real-time data to inform adaptive management. These innovations will make ecosystem-based mitigation more efficient, scalable, and transparent.

Research and development will continue to improve understanding of ecosystem processes and identify novel solutions for integrating nature-based approaches into economic systems. Open-access knowledge platforms will foster collaboration and accelerate the dissemination of best practices globally.

Robust Governance and Global Collaboration

Achieving a sustainable and resilient future requires strong governance frameworks and global collaboration. International agreements like the Paris Agreement and the Kunming-Montreal Global Biodiversity Framework will guide efforts to mainstream ecosystem-based mitigation into national policies and global climate strategies. Multi-stakeholder partnerships will unite governments, private sector actors, NGOs, and academic institutions to mobilize resources and coordinate efforts.

Transparent monitoring, reporting, and verification (MRV) systems will ensure accountability and demonstrate progress toward shared goals. Financial mechanisms, such as green bonds, carbon markets, and blended finance, will secure long-term funding for ecosystem-based initiatives.

A sustainable and resilient future envisions a world where ecosystems thrive, communities flourish, and humanity's impact on the planet is balanced with nature's capacity to regenerate. By prioritizing ecosystem-based approaches, leveraging innovation, and fostering inclusive governance, this vision can become a reality. Collective action and long-term commitment are essential to ensuring that the benefits of a sustainable future are shared by all, today and in the generations to come.

Conclusion

The conclusion brings together the key insights from this book, reinforcing the critical role of ecosystems in addressing the climate crisis. It highlights the interconnected themes of ecosystem-based mitigation, co-benefits for biodiversity and livelihoods, and the importance of overcoming barriers through innovation and collaboration. This chapter reflects on the progress made, acknowledges the challenges that remain, and sets a forward-looking vision for scaling nature-based solutions.

By providing a call to action for policymakers, practitioners, and individuals, the conclusion underscores the shared responsibility of safeguarding ecosystems as vital assets in achieving climate and sustainability goals. Together, these efforts pave the way for a resilient and sustainable future for the planet.

Recap of Key Themes Discussed in the Book

Ecosystem-based mitigation represents a transformative approach to addressing the dual crises of climate change and biodiversity loss while delivering socio-economic co-benefits. Throughout this book, the critical role of ecosystems as natural solutions for climate mitigation and adaptation has been explored, highlighting their capacity to sequester carbon, regulate ecosystems, and enhance resilience against climate impacts.

The initial chapters examined foundational concepts, including the definition, scope, and potential of ecosystem-based mitigation. Emphasis was placed on the interdependence of ecosystems and climate systems, showcasing how forests, wetlands, grasslands, and marine environments function as carbon sinks while supporting biodiversity and human well-being. The importance of co-benefits—such as improved livelihoods, water security, and disaster risk reduction—was underscored, demonstrating how ecosystem-based approaches can contribute to sustainable development.

Subsequent chapters delved into specific ecosystems, such as forests, peatlands, and coastal ecosystems, analyzing their unique roles in carbon sequestration and climate adaptation. Practical measures, including reforestation, agroforestry, and wetland restoration, were outlined as effective strategies for enhancing the capacity of these ecosystems. Policy frameworks, such as REDD+ and PES, were presented as tools to incentivize conservation and ensure financial sustainability.

The book also addressed challenges in implementation, including financial, technical, institutional, and social barriers. Strategies to overcome these obstacles were explored, emphasizing capacity building, community engagement, and multi-stakeholder collaboration. The role of technology, such as remote sensing and GIS, was highlighted as critical for monitoring progress and informing decision-making.

Finally, a forward-looking perspective was provided, examining emerging trends, innovations, and pathways for mainstreaming nature-based solutions into development planning. The vision for a sustainable and resilient future emphasized the need for ecosystems to be at the forefront of global climate strategies, supported by robust governance, inclusive policies, and long-term commitment.

Importance of Collaboration and Innovation in Ecosystem-Based Mitigation

Collaboration and innovation are cornerstones for advancing ecosystem-based mitigation at scale. The complex and interconnected challenges of climate change and ecosystem degradation cannot be addressed by any single actor or sector. Instead, coordinated efforts involving governments, private enterprises, communities, and international organizations are essential to ensure meaningful and sustainable outcomes.

Collaboration Across Sectors

Multi-stakeholder partnerships bring together diverse expertise, resources, and perspectives, fostering comprehensive solutions. Governments play a pivotal role in creating enabling policies and regulations, while the private sector contributes funding, technological innovation, and operational efficiency. NGOs act as intermediaries, ensuring that projects are inclusive and aligned with local needs. Academic institutions provide critical research and data to guide evidence-based decision-making.

Community involvement is central to successful implementation. Empowering local populations, particularly indigenous communities, ensures that solutions are culturally appropriate and context-specific. Inclusive approaches that integrate traditional knowledge with modern practices strengthen both the social and environmental impact of ecosystem-based mitigation.

Innovation as a Driver of Progress

Technological advancements have transformed the potential for ecosystem-based mitigation. Remote sensing technologies, IoT sensors, and artificial intelligence enable precise monitoring of ecosystems and carbon dynamics, enhancing the scalability and effectiveness of projects. Tools like GIS support spatial planning, ensuring that conservation efforts are targeted and efficient.

Innovative financial mechanisms, such as carbon markets, green bonds, and blended finance, mobilize resources and create incentives for sustainable practices. Blockchain technology enhances transparency and trust in carbon credit transactions, while digital platforms facilitate knowledge sharing and global collaboration.

Global Collaboration and Shared Accountability

International frameworks, such as the Paris Agreement and the Kunming-Montreal Global Biodiversity Framework, provide the foundation for coordinated global action. By aligning national strategies with international goals, countries can scale up ecosystem-based approaches and track progress through transparent monitoring, reporting, and verification systems.

Collaboration and innovation not only accelerate implementation but also build resilience against future challenges. Strengthening these pillars is essential for ensuring that ecosystem-based mitigation delivers its full potential in combating climate change, conserving biodiversity, and supporting sustainable development.

Call to Action for Policymakers, Practitioners, and Individuals

The path to a sustainable and resilient future requires collective action. Policymakers must prioritize ecosystem-based mitigation in national climate strategies, enacting policies that protect and restore natural systems while addressing socio-economic inequalities. Investments in nature-based solutions should be scaled up through innovative financial mechanisms, and collaborative frameworks should be strengthened to align efforts across sectors and borders.

Practitioners, including conservationists, scientists, and project managers, are encouraged to adopt cutting-edge technologies and data-driven approaches to optimize the design, implementation, and monitoring of ecosystem-based projects. Emphasizing community involvement and equitable benefit-sharing is critical for ensuring the longevity and success of these initiatives.

Individuals also have a vital role to play. Supporting ecosystem-friendly practices, advocating for sustainable policies, and participating in local conservation efforts contribute to global climate goals. Personal choices, such as reducing carbon footprints and supporting sustainable businesses, amplify collective impact.

Ecosystem-based mitigation is not just an environmental imperative—it is a shared responsibility and a pathway to a better future. By working together across all levels of society, we can harness the power of nature to mitigate climate change, safeguard biodiversity, and ensure a sustainable planet for generations to come.

www.ingramcontent.com/pod-product-compliance
Lightning Source LLC
Chambersburg PA
CBHW071601200326
41519CB00021BB/6833